Business Letters *for the*

CONSTRUCTION INDUSTRY

A Guide to Construction Communication

By Andrew Atkinson

BNi Building News

Editor-In-Chief
William D. Mahoney, P.E.

Technical Services
Sara Gustafson
Heidi Hassan
Vincent Wilhelm

Design
Robert O. Wright

BNi PUBLICATIONS, INC.

990 Park Center Drive, Ste E
Vista, CA 92081

LOS ANGELES
10801 National Blvd., Ste. 100
Los Angeles, CA 90064

ANAHEIM
1612 S. Clementine St.
Anaheim, CA 92802

1-800-873-6397

ISBN 978-1-55701-609-6

Table of Contents

Foreword

More than a book of letters, this a useful and practical tool every professional should have. Use these templates to expand into new markets and impress your clients by the efficiency of your communication and record keeping. *BNi Business Letters for the Construction Industry* is the perfect companion for *BNi Manual of Procedures & Form Book*.

Construction professionals are part of a project-based industry in which a system of accurate records is essential for survival. However, there has been a lack of attention to properly and accurately keep a written record of communication between clients, architects, subcontractors and trades people. This deficiency has resulted in costly problems that range from misunderstood winning bids to losing clients for lack of response. In some cases, the problem can escalate to litigation. Having the proper letter format and intent is critical to creating a successful system of communication involving all the participants of this complicated business.

This book is divided into four main chapters: *Letters to Clients, Letters to Field Professionals, Industry Support Letters, and Letters to Personnel.* The letter section of this book is compromised of two sheets per letter. The first page (even numbered page) describes the *Letter Type, Recipient, Letter Subject*, and *Letter Scenario*. It also offers a comprehensive list of *BNi-Form* references that may be used in relation to each scenario. The second page (odd numbered page) is the actual letter template. The letters are categorized by *Letter Type*. Each type covers a wide range of realistic scenarios that are exclusive to the construction industry. The main types can be described as follows:

APOLOGETIC:

Responsibility for and willingness to resolve the conflict at hand.

EXPLANATORY:

Describes a situation in a response to a particular request.

INFORMATORY:

Releases critical information that may require a request to take action.

JUSTIFICATION:

Describes and explains the writer's side of a situation.

REQUEST:

Requires the person receiving the letter to respond or take action.

CONFIRMATION:

Confirms points already agreed to.

CLARIFICATION:

Corrects erroneous or incomplete information.

AGREEMENT:

Corroborates information or approves contractual agreements.

CONGRATULATORY:

Rewards and recognizes a success or an individual accomplishment.

COURTESY:

Releases information in response to a particular request, and persuades the recipient to respond or take action.

CHANGE ORDER:

Summary of cost-related increases. Explains and justifies by persuading or demanding the recipient to take action.

Introduction

HOW TO USE THIS BOOK

Before you customize your letter, save the original template in a different location so that you always have the original templates unaffected by your editing. On the new letter template, simply turn on the *"Show Gridlines"* prompt under the *"Table"* prompt located on the upper screen of your *Microsoft Word* software. The table gridlines will appear and allow you to insert and edit information. You can also insert your company logo and letterhead by importing it as a text or image file into its allocated table. Always double check your letter for spelling and grammar mistakes. Once done, save the new letter on a dated file system where you can easily trace your correspondence.

Insert your Logo or Letterhead

Letter Description (Do Not Edit)   **Letter Body (Customized)**

LETTER TYPE: `APOLOGETIC`
ADDRESSED TO: **CLIENT**
RE: **BILLING ERROR**

SCENARIO:

A client for an ongoing building maintenance project has identified a billing mistake and placed a complaint regarding a specific invoice.

(Even Page Number)

COMPANY LETTERHEAD

(Date)

(Recipient's Name)
(Recipient's Title)
(Recipient's Contact Info.)

RE: (Project's Name and Tracking Number)
Dear (Recipient's Name),

Please accept our apologies regarding the overpayment noted on your most recent invoice in the amount of ($ amount). We have made the necessary adjustments to reflect the amount of ($ amount). We apologize for any inconvenience. We appreciate the input that customers like you give us, and look forward to doing business with you.

Respectfully,

(Name)
(Title)

Company Name, Address, Tel., Fax., E-mail, Web-Site Address

(Odd Page Number)

Insert your Company's Information

Introduction

COMMUNICATION PROTOCOL

The standard practice is very clear regarding construction communication. The owner communicates directly with the design principal (an architect, engineer, or construction manager), and through the design principal with the contractor. The owner does not communicate in writing directly with the construction inspector, consultants, subcontractor, vendors, or suppliers. Verbal communication between the owner and contractor should be avoided, except when circumstances require additional clarification, and then only in the presence of the design principal. Written communication should always go directly from the owner to the design principal, and from the design principal to the contractor.

The design principal communicates directly with the owner, contractor or more specifically, the contractor's superintendent, the design consultants, and the construction inspector. The design principal does not communicate directly with the subcontractors, or vendors, or suppliers. When circumstances require additional clarification, verbal communications may occur in the presence of the contractor. Written communications should be from the design principal to the contractor and from the contractor to the subcontractors, vendors, and suppliers.

The contractor communicates directly with the design principal and with the contractor's subcontractors, vendors, and suppliers. The contractor communicates with the owner through the design principal, and with the construction inspector through the contractor's superintendent. The contractor does not directly communicate in writing with the design consultants. Verbal communication between the contractor and the design consultants should be avoided except in the presence of the design principal and then only when circumstances require additional clarification.

The construction inspector communicates directly with the design principal and the contractor's superintendent. The inspector does not communicate in writing directly with the owner, design consultants, subcontractors, vendors, or suppliers.

The communication tool, the letter, begins and ends with the main office. Each office should establish standard procedures that will allow letters from all participants to be kept according to its designated project number. Establish a folder system on the early stages of each project. Identify each folder with the name or discipline of the participants (i.e. architect, concrete sub, etc.). Make sure you include electronic communication such as critical e-mails. The file system should be accessible through a network configuration that can be easily obtained from the jobsite's laptop or on a business trip.

PROTOCOL EXCEPTIONS

If the general contractor is acting as a design-build entity and is held responsible for errors of omissions, the contractor can assume the directive reserved for the design principal. This carries serious design liability, and the contractor should demonstrate the knowledge and experience required by designating a licensed design professional in charge of this directive (i.e. a licensed architect hired as the project architect for the job under the contractor's insurance umbrella if the professional is not insured). You should consult your attorney and insurance providers when setting up this type of arrangement.

When the construction inspector is retained by the owner as a permanent or staff employee, the inspector may communicate directly with the owner. In all such instances, the design principal should direct the construction inspector's work and be fully informed of communications concerning it.

The design principal may delegate direct communications between design consultants and construction inspectors or the contractor to expedite the work. Regardless of this delegation, the design principal must be fully informed in writing of all communications. The design principal must approve in writing any actions from communications directly from the direct contractor to the design consultants.

The contractor may specifically or partially delegate direct communications between his subcontractors, vendors, and suppliers and the design principal, design consultants, or the construction inspector. However, such design communications require both the contractor and design professional to be fully informed in writing of any communications, and each must approve in writing any resultant actions.

The owner communicates directly with the contractor concerning legal or contractual matters. However, the owner should keep the design principal informed in writing of all direct owner to contractor communications and any resulting actions.

Other direct communications between members of the design and construction teams may be made where required by laws or codes; however, all responsible parties should be kept informed in writing.

Introduction

COMMUNICATION COORDINATION

Written correspondence among the construction participants should be used along with the contract documents. Such correspondence is not as official as the agreement, contract documents, and change orders but is a necessary part of the process and is to be prepared with the same professionalism as the rest of the documents.

Use simple declarative statements and be clear in your intent. Each form of correspondence must include the project name, project number (especially where there might be more than one phase to a project or the potential of future phases), date, and the applicable parties to be involved with the correspondence.

The project name should always be consistent throughout the correspondence and the contract documents, preferably the same title used on the title sheet of the drawings. Where there is any possibility for confusion the name should include sufficient information to differentiate between multiple projects, such as including a building address, phase number, or similar location identifier. Project numbers can also be helpful in verification, although most projects have different numbers for each of the parties involved. For a more professional result, use your company's letterhead showing your logo and company information.

Date each piece of correspondence. Once a subject becomes important enough to require special correspondence, it is possible that the correspondence will result in various possible decisions. Where multiple correspondences exist on one subject, the dates on the correspondence identify which is the most recent decision. Identification of the applicable persons might be as simple as identifying who the correspondence is from and to whom is directed. It may also include an extensive list of persons who require the information in order to ensure proper coordination. This is frequently done with a "cc" list in the letter, fax, or e-mail. Blind copies might also be appropriate for certain types of letters. The decision to include blind copies should be based on prior mutual agreement between specific persons, such as the owner requesting a blind copy of correspondence between the construction manager and design principal.

Follow-up your phone conversations in writing for important matters as soon as possible, making sure the distribution list includes all affected parties. Enclose copies of correspondence received from subcontractors, suppliers, and vendors with a correspondence "cover" for all matters affecting the work. Coordinating the various disciplines during design is usually the responsibility of either the architect, principal engineer, or construction manager. Coordinating the work of various suppliers, trades, and subcontractors in the responsibility of the contractor as part of the construction means and methods. However, successful coordination requires that all parties of both the design and construction teams communicate with each other so that the construction can proceed in an orderly manner. Proper form and process need to be defined and followed to ensure that the information is properly transmitted and officially received. While speech is used extensively throughout the construction process through meetings and phone conversations, scheduling and legal procedures require written records with proper distribution to all concerned entities. Generally, letter coordination outlines the traditional communications procedures that have been recognized by the industry as *"protocol"* to keep the participants involved as fully informed as possible and to avoid misunderstandings. Naturally, exceptions occur. Letter communication also offers the opportunity to introduce standards, instructions and forms critical for a good record keeping system.

Introduction

Standard Terms

Addendum: Written or graphic instruments issued prior to the execution of the contract which modify or interpret the bidding documents, including drawings or specifications, by additions, deletions, clarifications, or corrections. Addenda will become part of the contract documents when the construction contract is executed. Plural *Addenda.*

Agency: Administrative subdivision of a public or private organization having jurisdiction over construction of the work.

Application and Certificate for Payment: Contractor's written request for payment of amount due for completed portions of the work, and, if the contract so provides, for materials delivered and suitably stored pending their incorporation into the work.

Accepted: Written acknowledgement of review by the architect/engineer or other authority having jurisdiction.

Architect: Designation reserved, usually by law, for a person or organization professionally qualified and duly licensed to perform architectural services, including analysis of project requirements, creation and development of project design, preparation of drawings, specifications and bidding requirements, and general administration of the construction contract. As used in this manual, the prime design professional with whom the owner contracts for design services: either the architect or the engineer.

Architect's Representative: An individual assigned by the architect to act as his liaison to assist in the administration of the construction contract.

Beneficial Occupancy: Use of a project or portion thereof for the purpose intended.

Building Inspector: A representative of a governmental authority employed to inspect construction for compliance with applicable codes, regulations, and ordinances.

Certificate for Payment: A statement from the architect to the owner confirming the amount of money due the contractor for work accomplished, or materials and equipment suitably stored, or both.

Change Order: A written order to the contractor signed by the owner and the architect, issued after the execution of the contract, authorizing a change in work or an adjustment in the contract sum or the contract time.

Codes: Regulations, ordinances, or statutory requirements of a governmental unit relating to building construction and occupancy, adopted and administered for the protection of the public health, safety, and welfare.

Consultant: An individual or organization engaged by the architect/engineer to render professional consulting services complementary to or supplementing his/her services.

Construction Documents: The owner-contractor agreement, the conditions of the contract (general, supplementary, and other conditions), the drawings, the specifications, and all addenda issued prior to execution of the contract, all modifications thereto, and any other items specifically stipulated as being included in the contract documents.

Construction Inspector: A qualified person engaged to provide full-time inspection of the work. In this manual, may refer to the inspector, specialized inspectors, and staff.

Introduction

Contract: The legally enforceable promise or agreement executed by the owner and the contractor for the construction of the work.

Contract Documents: The owner-contractor agreement, the conditions of the contract (general, supplementary, and other conditions), the drawings, the specifications, and all addenda issued prior to the execution of the contract, all modifications thereto, and any other items specifically stipulated as being included in the contract documents.

Contractor: The person or organization performing the work and identified as such in the contract.

Date of Substantial Completion: The date certified by the architect when the work or a designated portion thereof is sufficiently complete in substantial accordance with the contract documents so that the owner may occupy the work or designated portion thereof for the use for which it was intended.

Engineer: Designation reserved, usually by law, for a person or organization professionally qualified and duly licensed to perform engineering services, including analysis of project requirements, development of project design, preparation of drawings, specifications and bidding requirements, and general administration of the construction contract. See also *Consultant.*

Field Change Order: A written order used for emergency instruction to the contractor where the time required for the preparation and execution of a formal change order would result in a delay or stoppage of the work. The usage of a field change order may be subject to prior legal approval and limitations. A duly authorized change order replaces a field change order expediently.

Final Acceptance: The owner's acceptance of the project from the contractor upon certification by the architect that it is complete and in substantial accordance with the contract requirements. Final acceptance is confirmed by making of final payment unless otherwise stipulated at the time of making such payment.

Final Inspection: Final review of the project from the contractor upon certification by the architect that is complete and in substantial accordance with the contract requirements. Final acceptance is confirmed by making final payment unless otherwise stipulated at the time of making such payment.

Inspection: Examination of the work completed or in progress to determine its compliance with contract requirements. The architect ordinarily makes only two inspections of a construction project, one to determine the date of substantial completion, and the other to determine final completion. These inspections should be distinguished from the more general observations of visually exposed and accessible conditions made by the architect on periodic site visits during the progress of the work. A full-time construction inspector makes continuous inspections.

Inspection List: A list of items of work to be completed or corrected by the contractor.

Owner: (1) The architect's client and party to owner-architect agreement; (2) the owner of the project and party to the owner-contractor agreement.

Owner Representative: A person delegated by the owner to act in his/her behalf as liaison with the architect during the development of the project and construction of the work. The responsibilities delegated should be stipulated to the extent that this person may or can make decisions on the part of the owner.

Progress Payment: Partial payment made during the progress of the work on account of work completed and/or materials received and suitably stored.

Introduction

Progress Schedule: A diagram, graph, or other pictorial or written schedule showing proposed or actual times of starting and completion of the various elements of the work.

Project: The total construction designed by the architect, of which work performed under the contract documents may be a whole or part.

Punch List: See Inspection List.

Record Drawings: Construction drawings revised to show significant changes made during the construction process, usually based on marked-up prints, drawings, and other data furnished by the contractor to the architect. Sometimes the term "as-built drawings" has been used in the past. The contractor is responsible for the accuracy on the information provided.

Required: Need of contract documents, code, agency, normally accepted practice, or other authority unless context implies a different meaning.

Schedule of Values: A statement furnished by the contractor to the architect reflecting the portions of the contract sum allotted for the various parts of the whole and used as the basis for reviewing the contractor's applications for progress payments.

Shop Drawing: Drawings, diagrams, illustrations, schedules, performance charts, brochures, and other data prepared by the Contractor or any subcontractor, manufacturer, supplier, or distributor which illustrate how specific portions of the work shall be fabricated and/or installed.

Standards: Organizations or public agencies that are recognized, or have established by authority, custom general consent, industry standards, or other manner of a method, criterion example, or test for the manufacture, installation, workmanship, or performance of a material or system.

Subcontractor: A person or organization that has a direct contract with a prime contractor to perform a portion of the work at the site.

Substantial Completion: See *Date of Substantial Completion.*

Supplier: A person or organization that supplies materials or equipment for the work, including that fabricated to a special design, but does not perform labor at the site. See also *Vendor.*

Superintendent: Contractor's representative at the site who is responsible for continuous field supervision, coordination, completion of the work, and, unless another person is designated in writing by the contractor to the owner and architect, prevention of accidents.

Vendor: A person or organization that furnishes materials or equipment not fabricated to a special design for the work. See also *Supplier.*

Work: All labor necessary to produce the construction required by the contract documents, and all materials and equipment incorporated or to be incorporated in such construction.

Chapter One

Letters to Clients

This chapter explores the diverse business aspects of serving the client. In all letters, the best intent to provide customer service with the emphasis on conflict resolution. This section also affirms the professional attitude and tone the contractor requires addressing delicate and essential issues, such as payments and legal rights. These letters are extremely effective when used in combination with the forms listed below the *Letter Scenarios.*

LETTER TYPE: APOLOGETIC
ADDRESSED TO: CLIENT
RE: BILLING ERROR

SCENARIO:

A client for an ongoing building maintenance project has identified a billing mistake and filed a complaint regarding the invoice.

COMPANY LETTERHEAD

(Recipient's Name)   (Date)
(Recipient's Title)
(Recipient's Contact Info.)

RE: (Project's Name and Tracking Number)

Dear (Recipient's Name),

Please accept our apology regarding the overpayment noted on your most recent invoice in the amount of ($ amount).

We have made the necessary adjustments to reflect the amount of ($ amount). We apologize for any inconvenience. We appreciate the input customers like you give us and look forward to doing business with you in the future.

Respectfully,

(Name)
(Title)

Company Name, Address, Tel., Fax., E-mail, Web-Site Address

LETTER TYPE:	**APOLOGETIC**
ADDRESSED TO:	**CLIENT**
RE:	**INCORRECT INFORMATION RELEASE**

SCENARIO:

The project manager for a major communications provider requested cost-related information for his monthly review submittal to upper management.

The contractor released inaccurate information and realized he needs to provide clarification to maintain the client's trust.

COMPANY LETTERHEAD

(Recipient's Name)   (Date)
(Recipient's Title)
(Recipient's Contact Info.)

RE: (Project's Name and Tracking Number)

Dear (Recipient's Name),

It is my understanding that you received inaccurate information in the (date) cost report recently submitted to you.

Please note the following clarification: steel beams and trusses have increased in price due to the current market condition. Please also note that a five percent increase/adjustment needs to be made on all structural steel. The prices are expected to escalate even further for (year).

It is with regret that I must inform you about this unfortunate but necessary cost increase.

Please feel free to contact me if you need any further clarification.

Sincerely,

(Name)
(Title)

Company Name, Address, Tel., Fax., E-mail, Web-Site Address

LETTER TYPE:	APOLOGETIC
ADDRESSED TO:	**CLIENT**
RE:	**AUTHORIZATION FOR PAYMENT**

SCENARIO:

The contractor of a project has had payment withheld from a designated construction fund due to a stop notice. The matter has been cleared up, and the contractor needs to re-establish authorization for payment with the client and the construction fund.

REFERENCE NOTE:
Use-BNi-W Form 239 as the Authorization for Payment form if funds are retained on a construction fund.

COMPANY LETTERHEAD

(Recipient's Name)   (Date)
(Recipient's Title)
(Recipient's Contact Info.)

RE: (Project's Name and Tracking Number)

Dear (Recipient's Name),

It is with regret that I am forced to put a hold on the work at (address).

This decision was based on your financial institution's decision to issue a stop notice and discontinue payments. It is my understanding that this billing discrepancy has now been resolved and we are to resume construction.

I apologize for this unfortunate incident and assure you that you will continue to receive our services without further delay. We appreciate your business and look forward to conducting future business with you.

Cordially,

(Name)
(Title)

Company Name, Address, Tel., Fax., E-mail, Web-Site Address

LETTER TYPE:	**APOLOGETIC**
ADDRESSED TO:	**CLIENT**
RE:	**UNFINISHED WORK (PAYMENT)**

SCENARIO:

After several unsuccessful attempts by the contractor to cash check payments for a major residential remodel job, he decided to stop all work and write a lien notice to the client.

However, he opted to resolve this issue diplomatically in order to avoid a mechanic's lien. Fortunately, the client was unaware of the problem and intervened by resolving the problem with the bank. The contractor understood the client had no fault in this situation and decides to offer an apology.

COMPANY LETTERHEAD

(Recipient's Name)   (Date)
(Recipient's Title)
(Recipient's Contact Info.)

RE: (Project's Name and Tracking Number)

Dear (Recipient's Name),

    Please be advised that the reason we interrupted the schedule on the (name) project and stopped work was because your financial institution had issued stop payments on all of our billing.

    We understand that this situation has now been resolved and we will resume work immediately.

    We will submit a revised schedule for completion under separate cover within the next few days.

    Thank you for your understanding.

Sincerely,

(Name)
(Title)

Company Name, Address, Tel., Fax., E-mail, Web-Site Address

LETTER TYPE:	**EXPLANATORY**
ADDRESSED TO:	**CLIENT**
RE:	**JOINT CHECK AGREEMENT**

SCENARIO:

A client has heard of the *Joint Check Agreement* type of payment.

He is interested in using it for his project. However, he needs further clarification on how this process will work.

The contractor explains this process.

REFERENCE NOTE:
Use-BNi-W Form 330 as a Joint Check Agreement with the Owner and General Contractor
OR
Use-BNi-W Form 239 as the Joint Check Agreement with the Contractor and Subcontractor

---

COMPANY LETTERHEAD

(Recipient's Name)   (Date)
(Recipient's Title)
(Recipient's Contact Info.)

RE: (Project's Name and Tracking Number)

Dear (Recipient's Name),

I will be more than happy to discuss the *Joint Check Agreement* process with you.

This is an excellent way to provide protection from claims, liens, and stop notices by unpaid subcontractors, suppliers, or materialmen. Sometimes a joint check agreement is used in place of the requirement that the general contractor post a labor and material bond. With the joint check agreement, the owner agrees to issue joint checks made payable to the general contractor and to the subcontractor or materialman supplier for furnishing and supplying labor, material, services and equipment for your project. The general contractor, subcontractor, or materialman supplier in return, agree that each check issued and paid by you, the owner, shall be credited as payment. With the joint check agreement there is a 35-day limit from the day of the invoice.

Please let me know if you are still interested in this method of negotiation, and don't hesitate to contact me if you have additional questions.

Very truly yours,

(Name)
(Title)

Company Name, Address, Tel., Fax., E-mail, Web-Site Address

LETTER TYPE: **APOLOGETIC**
ADDRESSED TO: **CLIENT**
RE: **UNACCEPTABLE WORKMANSHIP
DUE TO CLIMATE**

SCENARIO:

The contractor for a major residential remodel has contracted to finish and close the project prior to the rainy season. Unfortunately, the wet season came earlier than anticipated, resulting in poor product performance. The exterior finishes such as varnishes and paint were damaged due to the humidity and rain.

COMPANY LETTERHEAD

(Recipient's Name)   (Date)
(Recipient's Title)
(Recipient's Contact Info.)

RE: (Project's Name and Tracking Number)

Dear (Recipient's Name),

Thank you for bringing this situation to my attention. The problem resulted from unexpected changes in the weather. The unanticipated wet season did not give us the opportunity for adequate drying on all of our exterior finishes prior to sealing.

Please know that I will take all necessary actions to resolve this problem by stripping and reapplying varnishes and paint. I will also make sure my crew provides temporary sheltering around and on top of such coatings.

Thank you very much for your patience and understanding.

Sincerely,

(Name)
(Title)

Company Name, Address, Tel., Fax., E-mail, Web-Site Address

LETTER TYPE:	**APOLOGETIC**
ADDRESSED TO:	**CLIENT**
RE:	**UNACCEPTABLE WORKMANSHIP (MATERIALS)**

SCENARIO:

The housing boom reduced the availability of lumber and in order to comply with a deadline, the framer made some bad business decisions. After a failed attempt to pass framing inspection, possible delays are of a concern to the client.

The general contractor decides to offer an apology.

> *REFERENCE NOTE:*
> *Use- BNi-W Form 312 as a Notice of Request for Change in Specifications and Substitutions.*

COMPANY LETTERHEAD

(Recipient's Name)   (Date)
(Recipient's Title)
(Recipient's Contact Info.)

RE: (Project's Name and Tracking Number)

Dear (Recipient's Name),

Please be aware that we have reviewed the framing inspector's notes. Even though we feel some of those items are not necessarily relevant, we will make all necessary adjustments and comply with the inspection. These items will be immediately addressed and resolved.

Please be assured that this unfortunate situation will not significantly delay our planned completion of your project.

Sincerely,

(Name)
(Title)

Company Name, Address, Tel., Fax., E-mail, Web-Site Address

LETTER TYPE:	**APOLOGETIC**
ADDRESSED TO:	**CLIENT**
RE:	**POOR WORKMANSHIP**
	(LABOR)

SCENARIO:

A fire-proofing subcontractor failed to attend basic installation training for a new fire-proofing product. The client noticed the poor condition of the installation during a scheduled walk-through. The issue was immediately brought to the attention of the general contractor. The contractor offers an apology.

COMPANY LETTERHEAD

(Recipient's Name)   (Date)
(Recipient's Title)
(Recipient's Contact Info.)

RE: (Project's Name and Tracking Number)

Dear (Recipient's Name),

Dear (Recipient's Name),

Thank you for bringing this problem to my attention. The deficiency resulted from inadequate training of the fire-proofing subcontractor. Recent fire codes regarding conduit fire-proofing require that all installers must attend proper training prior to installation. It appears that for this specific application and product, the sub failed to comply.

This matter will be dealt with. As your general contractor, I take full responsibility, and will provide an immediate trained replacement without any additional cost to you. We will demount all fire-proofing blocks. I will also try to make sure this will not significantly delay our original construction schedule.

Be assured that this situation does not reflect the typical craftsmanship and quality of my business operation. Thank you very much for your patience and understanding.

Sincerely,

(Name)
(Title)

Company Name, Address, Tel., Fax., E-mail, Web-Site Address

LETTER TYPE:	**APOLOGETIC**
ADDRESSED TO:	**CLIENT**
RE:	**DAMAGE OR LOSS OF PROPERTY**

SCENARIO:

The path of traffic on a partially occupied site was not completely cleared. The contractor provided instructions to the client and building's maintenance crew to clear all obstructions from the area of work. However, the building's maintenance crew ignored the notice and left personal property on the demolition site. The general contractor wants to apologize and compensate for any damage he is liable for through his insurance carrier.

COMPANY LETTERHEAD

(Recipient's Name)   (Date)
(Recipient's Title)
(Recipient's Contact Info.)

RE: (Project's Name and Tracking Number)

Dear (Recipient's Name),

Dear (Recipient's Name),

We acknowledge the problem at (address). Please take into consideration that the building's maintenance crew did not follow our noted instructions to clear all objects on the construction path staging area.

Please note that an insurance report has been filed. An agent of (company name) will contact you to discuss the course of action. Don't hesitate to reach me if you have any concerns.

Sincerely,

(Name)
(Title)

Company Name, Address, Tel., Fax., E-mail, Web-Site Address

LETTER TYPE:	**RELEASE OF INFORMATION**
ADDRESSED TO:	**CLIENT**
RE:	**WARRANTY**

SCENARIO:

The client of a kitchen remodel and family room addition project is extremely satisfied with the work performed. He congratulated the contractor on the results. He wants to receive the contractor's warranty in writing. The contractor responds.

REFERENCE NOTE:
Use- BNi-W Form 293 as the Contractor's Warranty Form

COMPANY LETTERHEAD

(Recipient's Name)   (Date)
(Recipient's Title)
(Recipient's Contact Info.)

RE: (Project's Name and Tracking Number)

Dear (Recipient's Name),

I am glad to hear that you are pleased with the work done on your home. Per your request, I am sending you a warranty. This warranty assures that our work is free from defects in material and workmanship for one year from the date of commencement, substantial completion, or notice of completion, whichever is first to occur. You should receive the written warranty and supporting manufacturer's warranties within five working days from the project's completion.

This Standard Limited Warranty is limited as follows:

- To the property only as long as it remains in the possession of the original owner named above.

- To the construction work that has not been subject to accident, misuse, or abuse.

- To the construction work that has not been modified, altered, defaced and/or had repairs made or attempted by others.

- That the contractor is immediately notified in writing within 10 days of first knowledge of defect by owner or his/her agents.

- That the contractor shall be given first opportunity to make any repairs, replacements and/or corrections to the defective construction at no cost to owner within a reasonable time.

- Under no circumstances shall the contractor be liable by virtue of this warranty or otherwise for damage to a person or property whatsoever for any special, indirect, secondary, or consequential damages of any nature arising out of the use or inability to use because of the construction defect.

Excluded from this warranty are materials and workmanship covered by warranties by others.

Thank you for your business. We look forward to serving your future construction needs and the needs of your friends and family.

(Name)

(Title)

Company Name, Address, Tel., Fax., E-mail, Web-Site Address

LETTER TYPE:	APOLOGETIC
ADDRESSED TO:	CLIENT
RE:	MISSING A DEADLINE
	(MATERIAL/DELIVERIES)

SCENARIO:

Due to an increase in commercial projects, demand for building materials has drastically increased. This situation has created a shortage of materials and affected their timely delivery to the jobsite. The developer is anxious to finish the project in order to meet the investor's open house deadline. The contractor realizes this situation is beyond his control and decides to offer an apology:

COMPANY LETTERHEAD

(Recipient's Name)   (Date)
(Recipient's Title)
(Recipient's Contact Info.)

RE: (Project's Name and Tracking Number)

Dear (Recipient's Name),

We had an agreement that the project would be finalized and a punch list released on (date) of this year.

Unfortunately, due to the high demand for construction materials, material deliveries are postponed until further notice. I am now forced to make some uncomfortable decisions regarding the completion date. However, I intend to resolve this matter by contacting additional manufacturers that may have the materials that we are looking for. We will get back to you as soon as possible with a new completion date.

If there is anything else I can do in order to maintain your trust, please don't hesitate to contact me.

I will be putting all my effort towards the speedy resolution of this problem.

Sincerely,

(Name)
(Title)

Company Name, Address, Tel., Fax., E-mail, Web-Site Address

LETTER TYPE: **COURTESY**
ADDRESSED TO: **CLIENT**
RE: **COURTESY FOLLOW-UP**

SCENARIO:

A remodeling contractor wants to thank a former client, while simultaneously taking the opportunity to market a new "green" service.

COMPANY LETTERHEAD

(Recipient's Name)   (Date)
(Recipient's Title)
(Recipient's Contact Info.)

RE: (Project's Name and Tracking Number)

Dear (Recipient's Name),

   I am glad to hear that you are pleased with your newly remodeled house. We know that you can look back and remember the great adventure it had been to transform your ideas from dreams into reality, it has been exciting.

   Our company is embarking on a major remodel project involving 2,500 square feet of prime entertainment space for a well-known local businessman and are offering a new variety of "green" products in response to the client's needs.

   We certainly enjoyed serving you, and we look forward to assisting you with any renovation project you may have in the future. Please let us know if you have any suggestions that could help us improve. Thank you very much for making this project a memorable one.

Very truly yours,

(Name)
(Title)

Company Name, Address, Tel., Fax., E-mail, Web-Site Address

LETTER TYPE: **CHANGE ORDER**
ADDRESSED TO: **CLIENT**
RE: **RESPONSIBILITY DETERMINATION**

SCENARIO:

A contractor writes letter to the owner regarding obligation to determine responsibility for questionable work.

REFERENCE NOTE:
Use-BNi-W Form 308 for Transmittal of Signed Subcontract
OR
Use-BNi-W Form 296 for Letter of Protest and Objection

COMPANY LETTERHEAD

(Recipient's Name)   (Date)
(Recipient's Title)
(Recipient's Contact Info.)

RE: (Project's Name and Tracking Number)

Dear (Recipient's Name),

Per our conversation on (date) both (Contractor) and (Subcontractor) need clarification of the responsibility for (brief description of questionable work). We consider this work to be outside the scope of our original contract.

General Conditions Article (number) directs that the owner (or architect, as described) interpret the requirements of the contract. Accordingly, please identify the party responsible for the subject work, indicating the applicable specification section and responsible contractor (subcontractor) by name.

Your complete written response is required by (date) to maintain job progress.

Very truly yours,

(Name)

(Title)

Company Name, Address, Tel., Fax., E-mail, Web-Site Address

LETTER TYPE: **CONFLICT OF INTEREST**
ADDRESSED TO: **CLIENT**
RE: **UNAUTHORIZED USE OF PERSONNEL**

SCENARIO:

A general contractor had a service contract for a property management client. The work had been going smoothly until the end of the second year. At that time, members of the management company approached the contractor's workers to perform jobs for a lower cost, without the permission of the contractor. The Contractor feels betrayed and addresses the situation:

COMPANY LETTERHEAD

(Recipient's Name)   (Date)
(Recipient's Title)
(Recipient's Contact Info.)

RE: (Project's Name and Tracking Number)

Dear (Recipient's Name),

We hope the services provided to your company have met your expectations.

We recently came upon some news that needs to be addressed. Our maintenance crew reported that they have been approached by members of your company to service additional buildings that are part of your client portfolio.

I have asked them to decline the offers for the following reasons: First, our crew members have an exclusive agreement to work for us and are under contract and are covered by our worker's compensation insurance policies. Second, we can't afford to expose ourselves to additional liability by covering work we do not supervise directly. Third and final, we can't offer you any type of warranty for worked performed.

I assume you were not aware of this situation. Please don't hesitate to contact me if you have any questions. Thank you for your continued support.

Sincerely,

(Name)
(Title)

Company Name, Address, Tel., Fax., E-mail, Web-Site Address

LETTER TYPE: **CHANGE ORDER**
ADDRESSED TO: **CLIENT**
RE: **DESIGN-DUPLICATIONS**

SCENARIO:

Contractor writes a letter to the owner regarding design duplications.

COMPANY LETTERHEAD

(Recipient's Name)   (Date)
(Recipient's Title)
(Recipient's Contact Info.)

RE: (Project's Name and Tracking Number)

Dear (Recipient's Name),

Section (number) requires the (type of contractor, i.e. concrete) contractor to provide the welded portion of the masonry anchors attached to steel columns. Detail (number) indicates the same work to be performed by the (type of contractor) contractor. Both contractors have refused to perform the work as part of their contracts.

General Conditions, Article (number), states that the specification takes precedence over information on the plans. Accordingly, please confirm that it is your intention to have the (type of contractor) contractor complete the work as part of Section (number).

Very truly yours,

(Name)

(Title)

Company Name, Address, Tel., Fax., E-mail, Web-Site Address

LETTER TYPE: **CHANGE ORDER**
ADDRESSED TO: **CLIENT**
RE: **OVERTIME COSTS**

SCENARIO:

The contractor for a project has been requested to work overtime by the client. The effort to finish phases of the project ahead of schedule was triggered by FEMA incentives to service flood victims. The contractor releases a change order to recuperate the expenses resulting from the excessive overtime. The following cover letter precedes the actual change order form:

REFERENCE NOTE:
Use-BNi-W Form 232 as the Change Order Summary if a quotation is requested for a change in scope of work under the contract or if claiming extra work or overtime.

COMPANY LETTERHEAD

(Recipient's Name)   (Date)
(Recipient's Title)
(Recipient's Contact Info.)

RE: (Project's Name and Tracking Number)

Dear (Recipient's Name),

Per our recent conversation regarding the extra labor costs incurred on your project, please find the enclosed Change Order Summary dated (date).

If the client requires changed or accelerated work which necessitates overtime, contractors on firm-price contracts must be fully compensated for loss of productivity, overtime premiums, and additional supervisory and administrative costs resulting from such overtime.

Please don't hesitate to contact me if you have any questions regarding the Change Order Summary, or if you find it necessary to meet in person to discuss and review each item listed on the form attached. Thank you for your support and understanding.

Very truly yours,

(Name)
(Title)

Company Name, Address, Tel., Fax., E-mail, Web-Site Address

LETTER TYPE: **CHANGE ORDER**
ADDRESSED TO: **CLIENT**
RE: **ADDED COSTS**

SCENARIO:

The contractor for a project has issued several change orders. The client is under the impression that the change orders appear to be items not covered under the original scope of work and assumes that a new contractual agreement is needed. The contractor clarifies that the items are part of the original agreement:

REFERENCE NOTE:
Use-BNi-W Form 211 as the Change Order form of change order for use in the performance of the contract.

COMPANY LETTERHEAD

(Recipient's Name)   (Date)
(Recipient's Title)
(Recipient's Contact Info.)

RE: (Project's Name and Tracking Number)

Dear (Recipient's Name),

In response to your concern regarding the change orders for your project, we would like to clarify that this is not a new agreement.

The items listed on the attached Change Order Summary enumerate changes necessary for the completion of your project. All provisions of the original contract documents shall remain unchanged except as specifically modified by the change order. The attached document is made with the reservation of a claim for attendant field costs, general administration costs, ripple effect costs, and extension of time.

Unless written objection to this change order is received within ten days, acceptance of this change order shall be effective by all parties. Please feel free to contact me if you have any questions regarding any of the items listed in the change orders or if you find it necessary to meet so that we may review the documents. Thank you for your cooperation.

Very truly yours,

(Name)
(Title)

Company Name, Address, Tel., Fax., E-mail, Web-Site Address

LETTER TYPE: REQUEST
ADDRESSED TO: CLIENT
RE: ALLOWANCE

SCENARIO:

The contractor of a theatre building has been requested by the client to apply an allowance stipulated on the original contract. The allowance is to be used for the upgrade of interior specialties such as expensive curtain systems and highly rated acoustical ceiling deflectors. The upgrades exceed the allotted allowance, and the contractor needs to explain the situation to the client to determine if the client is willing to pay the difference.

The contractor decides to write a letter to review the allowance:

REFERENCE NOTE:
Use-BNi-W Form 320 if you need to obtain a Notice of Directive or communication from the client or design professional.

COMPANY LETTERHEAD

(Recipient's Name)   (Date)
(Recipient's Title)
(Recipient's Contact Info.)

RE: (Project's Name and Tracking Number)

Dear (Recipient's Name),

I understand your concern regarding the applicability of the allowances established in our original contract. The products you have requested through your architect exceed the allowance amount designated. Per our agreement, when the contract price includes allowances and the cost of performing the job covered by the allowance is greater or less than the allowance, then the contract price shall be increased or decreased accordingly.

Unless otherwise requested by the owner in writing, the contractor shall use his own judgment in accomplishing work covered by an allowance. If the owner requests that work covered by an allowance be accomplished in such a way that the cost will exceed the allowance, the contractor shall comply with the owner's request, provided that the owner pays the additional cost. This, in brief, describes our contractual allowance agreement. In this situation, we are willing to follow your request for an upgrade on the interior specialties you requested.

However, we would need to recover from you the extra cost. Please let me know if you still want to proceed so that we may expedite your order by faxing you an invoice for the cost difference. Please contact me at your earliest convenience so that we may determine your direction. Thank you for your time and assistance in this matter.

Very truly yours,

(Name)
(Title)

Company Name, Address, Tel., Fax., E-mail, Web-Site Address

LETTER TYPE: CHANGE ORDER
ADDRESSED TO: CLIENT
RE: RFI APPROVAL

SCENARIO:

Contractor writes a letter to the owner regarding a change order approval on work not covered by the plans.

REFERENCE NOTE:
Use-BNi-W Form 277 for Notice of Claim for Extra Work
OR
Use-BNi-W Form 103 for Extra Work Orders
OR
Use-BNi-W Form 266 if you need to justify a payment by assignment.

COMPANY LETTERHEAD

(Recipient's Name)   (Date)
(Recipient's Title)
(Recipient's Contact Info.)

RE: (Project's Name and Tracking Number)

Dear (Recipient's Name),

Section (number) of pre-filed sub bid (number) indicates that the concrete encasement at the underground electrical ducts is to be installed *"in accordance with the requirements of Section 03300."* Neither the site plans nor Section 03300 incorporates this work.

Accordingly, we are requesting your approval of a change order in the amount of (amount) to complete this work. A detailed proposal with all substantiating documentation is attached for your review.

Your approval is required by (date). If we do not receive a response by this date, the progress of the project will be delayed and we will be required to add the costs attributable to the delay to the amount of this change order.

Very truly yours,

(Name)
(Title)

Company Name, Address, Tel., Fax., E-mail, Web-Site Address

LETTER TYPE: **CHANGE ORDER**
ADDRESSED TO: **CLIENT**
RE: **EQUIPMENT COORDINATION**

SCENARIO:

Contractor writes letter to the owner regarding owner-furnished equipment.

REFERENCE NOTE:
Use-BNi-W Form 205 for Release of responsibility for work not performed by contractor or for materials and equipment not furnished by contractor.

COMPANY LETTERHEAD

(Recipient's Name)   (Date)
(Recipient's Title)
(Recipient's Contact Info.)

RE: (Project's Name and Tracking Number)

Dear (Recipient's Name),

Please provide me with all information regarding owner-furnished equipment and material so that this equipment can be properly coordinated with the work.

We need this information, as well as the equipment and material, in sufficient time to allow incorporation without disruption or interference. Your attention is directed to the specific requirements set forth in the current construction schedule.

The construction is progressing in accordance with the current contract documents. We request your immediate response to minimize the impact of any possible change.

Thank you for your prompt response.

Very truly yours,

(Name)
(Title)

Company Name, Address, Tel., Fax., E-mail, Web-Site Address

LETTER TYPE:	**CHANGE ORDER**
ADDRESSED TO:	**CLIENT**
RE:	**EQUAL TO PROPRIETARY ITEM-1**

SCENARIO:

The client of a project has requested a credit for a product substitution.

This is a common result of a cost-plus contract where the contractor is managing the cost of materials and supplies. In this situation, the client is attempting to obtain a credit from the contractor by replacing a product with a substandard quality, non-equal one which will affect the contractor's liability and warranties. The contractor rejects the substitution.

> *REFERENCE NOTE:*
> *Use-BNi-W Form 292 if a back-charge will be implied*
> *OR*
> *Use-BNi-W Form 312 for Notice of Request for changes in Specifications and Substitutions.*

COMPANY LETTERHEAD

(Recipient's Name)   (Date)
(Recipient's Title)
(Recipient's Contact Info.)

RE: (Project's Name and Tracking Number)

Dear (Recipient's Name),

The (subject item) was submitted for approval in lieu of a previously approved specification.

Your decision to substitute this item on the condition of a credit change order is not in accordance with the terms of the contract and, as such, cannot be accepted.

We will be installing the original product to conform to the function, performance, and quality criteria of the specification, the use of this original product had been previously approved.

Very truly yours,

(Name)
(Title)

Company Name, Address, Tel., Fax., E-mail, Web-Site Address

LETTER TYPE: **CHANGE ORDER**
ADDRESSED TO: **CLIENT**
RE: **EQUAL TO PROPRIETARY ITEM-2**

SCENARIO:

Letter to the owner regarding the rejection of equal or proprietary item.

This is a common result of a labor-only contract where the client is managing the cost of materials and supplies.

In this situation, the client is attempting to save money by purchasing a sub-standard product that does not meet the specs and plans.

The contractor has to reject the substandard product since, it will increase his liability. In this particular scenario, the contractor is ordering the supplies through the client's account. The contractor has identified a competitive product that meets the project's requirements and needs to obtain the client's written authorization through a change order.

REFERENCE NOTE:
Use-BNi-W Form 292 if a back-charge will be implied
OR
Use-BNi-W Form 312 for Notice of Request for changes in Specifications and Substitutions.

COMPANY LETTERHEAD

(Recipient's Name)   (Date)
(Recipient's Title)
(Recipient's Contact Info.)

RE: (Project's Name and Tracking Number)

Dear (Recipient's Name),

The (substituted product) was submitted for approval in lieu of a previously approved specification. For the reasons stipulated in (refer to submission correspondence) the (substituted product) does not meet the specified requirements. Your decision to use the unspecified item (material) is not in accordance with the contract and, as such, cannot be accepted as equal.

Accordingly, the (name the specified item) will be provided for the additional sum of (amount) per detailed change order proposal, attached.

Your approval of this change order is required by (date). Please be advised that your failure to respond by this date constitutes acceptance of the change order by default. Any subsequent action which results in additional costs due to project delays and interferences will be added to this change order price. Please take notice.

Very truly yours,

(Name)
(Title)

Company Name, Address, Tel., Fax., E-mail, Web-Site Address

LETTER TYPE:	**CHANGE ORDER**
ADDRESSED TO:	**CLIENT**
RE:	**CHANGED SITE CONDITIONS**

SCENARIO:

Contractor writes letter to the owner regarding changed site conditions.

REFERENCE NOTE:
Use-BNi-W Form 211 as a change order form for use in the performance of the contract.

OR

Use-BNi-W Form 232 as a change order form if the client requires a quotation for a change in scope of work.

OR

Use-BNi-W Form 226 if you need to give official notice of a possible cost increase confirmation.

COMPANY LETTERHEAD

(Recipient's Name)   (Date)
(Recipient's Title)
(Recipient's Contact Info.)

RE: (Project's Name and Tracking Number)

Dear (Recipient's Name),

   Drawing (number) indicates the pipe invert elevation at (location) to be (value). The actual invert elevation is (value). The corresponding increase in the contract price is (amount), computed in accordance with the procedures of (article or section number). Refer to the detailed change order proposal attached.

   Your approval is required by (date) to allow the work to proceed without additional interruption. The effects of the change on the construction schedule are now being reviewed. When the analysis is complete, you will be notified of the additional costs.

   Please be aware that we reserve the right to claim all damages resulting from effects which we cannot anticipate at this time.

Very truly yours,

(Name)
(Title)

Company Name, Address, Tel., Fax., E-mail, Web-Site Address

LETTER TYPE: CHANGE ORDER
ADDRESSED TO: CLIENT
RE: CHANGED SITE CONDITIONS

SCENARIO:

Contractor writes a letter to the client for site changes that resulted from discrepancies between the drawings and field conditions.

COMPANY LETTERHEAD

(Recipient's Name)   (Date)
(Recipient's Title)
(Recipient's Contact Info.)

RE: (Project's Name and Tracking Number)

Dear (Recipient's Name),

It has come to my attention that the existing site grades are significantly different from those indicated on drawing (number). In response to my conversation with (name), we have made arrangements with a registered land surveyor to provide a corrected layout and calculation. When the data is complete, you will be advised of all costs associated with this change.

We anticipate that the survey will be complete on (date). Please confirm whether we should either stop work or proceed with the additional work by (date) to minimize the interferences caused by this change.

The effects of the change on the construction schedule are now being reviewed. When this analysis is complete, you will be notified of any additional costs.

Please be aware that we reserve the right to claim all damages resulting from effects which we cannot anticipate at this time.

Very truly yours,

(Name)
(Title)

Company Name, Address, Tel., Fax., E-mail, Web-Site Address

LETTER TYPE: CHANGE ORDER
ADDRESSED TO: CLIENT
RE: PENDING CHANGE ORDER-1

SCENARIO:

Contractor writes a letter to the client regarding a pending change order.

REFERENCE NOTE:
Use-BNi-W Form 211 as a Change Order Form for use in the performance of the contract.

OR

Use-BNi-W Form 232 as a Change Order Form if the client requires a quotation for a change in scope of work.

OR

Use-BNi-W Form 226 if you need to give Official Notice of a possible cost increase confirmation.

OR

Use-BNi-W Form 230 if you need to give a Notice of Excusable Delay and request for extension time

COMPANY LETTERHEAD

(Recipient's Name)   (Date)
(Recipient's Title)
(Recipient's Contact Info.)

RE: (Project's Name and Tracking Number)

Dear (Recipient's Name),

As reviewed at the job meeting on (date), the welded portion of the masonry anchors is not specified in either Section (number) Section (number).

At this time, it appears that there will be an increase in the contract sum to cover the cost of this additional work, and possibly a delay as well. We are proceeding with the assembly of all component prices and will produce a detailed analysis of all effects. Upon the completion of our analysis, a change order proposal will be submitted.

Please be advised that letter serves as your notification in accordance with the requirements of General Conditions, Article (number).

Very truly yours,

(Name)
(Title)

Company Name, Address, Tel., Fax., E-mail, Web-Site Address

LETTER TYPE: **CHANGE ORDER**
ADDRESSED TO: **CLIENT**
RE: **PENDING CHANGE ORDER-2**

SCENARIO:

Contractor writes letter to the client regarding a pending change order.

REFERENCE NOTE:
Use-BNi-W Form 211 as a Change Order Form for use in the performance of the contract.
OR
Use-BNi-W Form 232 as a Change Order Form if the client requires a quotation for a change in scope of work.
OR
Use-BNi-W Form 226 if you need to give Official Notice of a possible cost increase confirmation.
OR
Use-BNi-W Form 230 if you need to give a Notice of Excusable Delay and request for extension time

---

COMPANY LETTERHEAD

(Recipient's Name)   (Date)
(Recipient's Title)
(Recipient's Contact Info.)

RE: (Project's Name and Tracking Number)

Dear (Recipient's Name),

Per (refer to pending change order), our total price to perform the subject change is $( ). All supporting documentation is attached.

Our schedule analysis has determined that the resulting change will add (number) working days to the project schedule. This cost is included in the above price.

Your approval is required by (date) to maintain the above price and schedule; any approval after this date will increase the cost and time associated with this change, additional cost and time will be added to our proposal.

We call your attention to the exclusions detailed in the contractor's quotations regarding situations beyond the contractor's reasonable control.

Very truly yours,

(Name)
(Title)

Company Name, Address, Tel., Fax., E-mail, Web-Site Address

LETTER TYPE: **CHANGE ORDER**
ADDRESSED TO: **CLIENT**
RE: **COST ESCALATION**

SCENARIO:

Contractor writes a letter to the client regarding a daily accrued cost.

REFERENCE NOTE:
Use-BNi-W Form 321 as a notice of reserving impact costs

OR

Use-BNi-W Form 295 as a Change Order/Progress Payment Signing under Protest.

COMPANY LETTERHEAD

(Recipient's Name)   (Date)
(Recipient's Title)
(Recipient's Contact Info.)

RE: (Project's Name and Tracking Number)

Dear (Recipient's Name),

Change proposal (number), dated (date) for the subject work required your approval by (date). To date, we have not received any response to this proposal, and the lack of a definitive response is now creating additional costs that were not included in the original change order price.

Each day beyond the required approval date is adding ($ amount) per day to the change order price in administrative overhead alone. Additional items may also be added to the proposal price after your authorization to proceed is finally received. Accordingly, upon receipt of your approval, all costs resulting from the effects of the untimely response will be submitted to correct the final change order price.

We reserve the right to claim all damages resulting from any untimely approval actions.

Very truly yours,

(Name)
(Title)

Company Name, Address, Tel., Fax., E-mail, Web-Site Address

LETTER TYPE: <mark>CHANGE ORDER</mark>
ADDRESSED TO: CLIENT
RE: ACKNOWLEDGEMENT OF WORK PERFORMED

SCENARIO:

Contractor writes a letter to the client to acknowledge the performance of disputed work. The owner does not accept responsibility for the charges.

REFERENCE NOTE:
Use-BNi-W Form 321 as a Notice of Reserving Impact Costs
OR
Use-BNi-W Form 295 as a Change Order/Progress Payment Signing under Protest.
OR
Use-BNi-W Form 278 for Claim Sheet Daily Record of Labor, Equipment, and Materials.

COMPANY LETTERHEAD

(Recipient's Name)   (Date)
(Recipient's Title)
(Recipient's Contact Info.)

RE: (Project's Name and Tracking Number)

Dear (Recipient's Name),

    Per your request on (date), we are proceeding with the subject work under protest, in the interest of job progress.

    We will be preparing time and material tickets on a daily basis to document the actual work performed, along with all resources used. They will be presented to your on-site representative for signature at the end of each day.

    We are writing this letter to advise you of our position with regard to this work and to confirm that your representative's signature on the time and material tickets will only acknowledge the accuracy of information contained in the respective tickets, and will not indicate your acceptance of responsibility for the work at this time.

Very truly yours,

(Name)
(Title)

Company Name, Address, Tel., Fax., E-mail, Web-Site Address

LETTER TYPE: **CHANGE ORDER**
ADDRESSED TO: **CLIENT**
RE: **SPECIAL MEETING INVITATION**

SCENARIO:

Contractor writes a letter to the client regarding a meeting to discuss a change order.

REFERENCE NOTE:
Use-BNi-W Form 203 for Speed Memos when communicating with other firms and organizations.
                     OR
Use-BNi-W Form 203A for internal Speed Memos if you need a record reminder of the meeting.
                     OR
Use-BNi-W Form 212 for final Meeting Minutes Reporting.

COMPANY LETTERHEAD

(Recipient's Name)   (Date)
(Recipient's Title)
(Recipient's Contact Info.)

RE: (Project's Name and Tracking Number)

Dear (Recipient's Name),

Per our conversation on (date), a special meeting will be held at the jobsite on (date) to resolve the subject change. Each issue of the following agenda will be reviewed:

1.   (Description)

2.   (Description)

3.   (Description)

The following individuals (insert the names of each person required) are required to be present to discuss and resolve these issues.

Thank you for your consideration.

Very truly yours,

(Name)
(Title)

Company Name, Address, Tel., Fax., E-mail, Web-Site Address

LETTER TYPE:	**CHANGE ORDER**
ADDRESSED TO:	**CLIENT**
RE:	**EXTRA WORK CONFIRMATION**

SCENARIO:

Contractor writes a letter to the owner regarding a change order approval on work not covered by the plans.

REFERENCE NOTE:
Use-BNi-W Form 211 as a Change Order Form for use in the performance of the contract.
                    OR
Use-BNi-W Form 232 as a Change Order Form if the client requires a quotation for a change in scope of work.
                    OR
Use-BNi-W Form 226 if you need to give official notice of a possible cost increase confirmation.
                    OR
Use-BNi-W Form 230 if you need to give a Notice of Excusable Delay and request for extension time

---

COMPANY LETTERHEAD

(Recipient's Name)   (Date)
(Recipient's Title)
(Recipient's Contact Info.)

RE: (Project's Name and Tracking Number)

Dear (Recipients Name),

This letter confirms the conversation held between (contractor) and (owner) on (date) wherein we were instructed to perform the following work on the above described project:

1. (description of work)

2. (description of work)

3. (description of work)

The scope of work described above was not included in our original contract and is an "extra" that will be added under the payment dates described on the contract payment schedule. The time of payment for the extra work shall be as provided for in the contract.

For the purpose of this extra work order only, the term "cost" is defined to include the actual cost less any discounts allowed on all subcontracts, material, and direct labor used on the extra work including payroll taxes health and welfare and vacation fund contributions, workers' compensation, and other insurance premiums which are measured by payroll, sales taxes, cartage, equipment rental, and other direct costs incurred in performing the extra work. The term "cost" excludes supervisory work, other than working foremen, rent, utilities, and transportation provided by vehicles owned by the contractor.

Please let me know immediately if this memorandum fails in any way to confirm your understanding of the conversation.

Very truly yours,

(Name)
(Title)

Company Name, Address, Tel., Fax., E-mail, Web-Site Address

LETTER TYPE: CONFIRMATION
ADDRESSED TO: CLIENT
RE: DUTIES AND RESPONSIBILITIES

SCENARIO:

A client has requested that the contractor describe the project management tasks for his new retail building. The contractor establishes his involvement:

REFERENCE NOTE:
Use-BNi-W Form 203 for Speed Memos when communicating with other firms and organizations.

OR

Use-BNi-W Form 203A for internal Speed Memos if you need a record reminder of the meeting.

OR

Use-BNi-W Form 212 for final Meeting Minutes Reporting.

COMPANY LETTERHEAD

(Recipient's Name)   (Date)
(Recipient's Title)
(Recipient's Contact Info.)

RE: (Project's Name and Tracking Number)

Dear (Recipients Name),

In response to your request, we will conduct a pre-construction meeting to discuss our involvement. Our main management task will be the coordination of subcontractor work, change order records, periodic site inspections, drawing/specification follow-through, and correction of any discrepancies with the architect and subcontractors. Our office will prepare applications for payment to subcontractors in accordance with payment schedules (except for final payment). Prior to making final payment to subs, we will establish that all work is complete and satisfactory.

We will ensure inspection upon completion by developing a punch list prior to making final payments to subs. Waiver of lien certificates will be obtained from subs before project completion. Upon final payment release, we will deliver the following documents to you:

1. Equipment warranty & operating instructions

2. Certificates of inspection

3. Occupancy certificate

4. Contractor's release of liens

Please feel free to contact me if you have any additional questions or concerns prior to our meeting. An invitation will be sent to you shortly detailing the date and time. We look forward to serving your construction management needs.

Respectfully,

(Name)

(Title)

Company Name, Address, Tel., Fax., E-mail, Web-Site Address

LETTER TYPE: **CONFIRMATION**
ADDRESSED TO: **CLIENT**
RE: **DESIGN PHASE REQUIREMENTS**

SCENARIO:

A design-build contractor informs a client of the design phase requirements of a storage building project.

REFERENCE NOTE:
Use-BNi-W Form 203 for Speed Memos when communicating with other firms and organizations.

OR

Use-BNi-W Form 203A for internal Speed Memos if you need a record reminder of the meeting.

OR

Use-BNi-W Form 212 for final Meeting Minutes Reporting.

COMPANY LETTERHEAD

(Recipient's Name)   (Date)
(Recipient's Title)
(Recipient's Contact Info.)

RE: (Project's Name and Tracking Number)

Dear (Recipients Name),

As a design-build firm, we look to establish a process for design review and changes. This relates to any items provided by you as the owner in order to determine any special considerations regarding the zoning and environmental impact for this sensitive site. The initial phase will help us establish project files and records, as well as consulting with the local building inspector, if required, in order to determine specific project requirements. As a service to you, we will negotiate variances in order to obtain complete or partial mitigation of your site. The pre-design phase is also instrumental in helping us determine permit approval.

We must also obtain a land survey to determine the boundaries of your lot. This information will be critical in assessing the proximity and availability of utilities as well as any special geographical features that could result in infrastructure impediments. At that point we will determine if additional professional services are required, such as a land surveyor or a civil engineer. As a design-build contractor, we will assist you with this professional selection process.

All associated permit fees and reimbursable expense items will be treated as specified on the original contract. Additional professional services will be accorded per your acceptance, and will be set up for direct billing, unless you want us to mediate services and payments. For transparency and record keeping, we suggest separate contracts and billing.

We take pride in serving your construction needs. Please don't hesitate to contact me if you have any questions or concerns regarding this or any other phase of your project.

Best Regards,

(Name)
(Title)

Company Name, Address, Tel., Fax., E-mail, Web-Site Address

LETTER TYPE: CONFIRMATION
ADDRESSED TO: CLIENT
RE: PROPERTY SALE CONFIRMATION

SCENARIO:

A contractor-investor is selling a house. After receiving final offers, he decides to reduce the price if the following conditions are established.

REFERENCE NOTE:
Use-BNi-W Form 271 for Purchase "AS-IS" Agreement Personal Property.
OR
Use-BNi-W Form 271A for Purchase "AS IS" Agreement Real Property.

COMPANY LETTERHEAD

(Recipient's Name)   (Date)
(Recipient's Title)
(Recipient's Contact Info.)

RE: (Project's Name and Tracking Number)

Dear (Recipients Name),

Per your request regarding the property described above, I am considering a price reduction under the following conditional agreement:

1. That "AS IS" shall include any and all conditions, known or unknown, anticipated or unanticipated.
2. That upon advice of legal counsel, the undersigned hereby waives and releases all rights as to "sellers", herein.
3. That sellers and/or their agents have made no representations whatsoever, except title, to buyers.
4. That sellers have requested buyers to make and to have their own independent investigations of the property.
5. That buyers have made their own independent investigations, examinations, and inspections of the subject property and rely solely on their own investigation, examination and inspections.
6. That the property is sold and purchased in "AS IS" condition, without any warranty expressed or implied whatsoever, and buyer assumes all and any responsibility for the condition of the property and hereby unconditionally waives and releases and forever discharges sellers of and from all claims, demands, actions and causes of action arising out of or in any way connected with the sale purchase.

The terms and conditions of the purchase described above will be included in the contract and will require your written approval.

Please let me know if you have any questions or concerns. Thank you for your interest in my property.

Very truly yours,

(Name)
(Title)

Company Name, Address, Tel., Fax., E-mail, Web-Site Address

LETTER TYPE: **CONFIRMATION**
ADDRESSED TO: **CLIENT**
RE: **MANUFACTURER'S REPRESENTATIVES**

SCENARIO:

The client for a building project requests that the contractor provide all equipment warranties for staff review. The contractor explains that the warranties will be provided after proper testing and inspection.

COMPANY LETTERHEAD

(Recipient's Name)   (Date)
(Recipient's Title)
(Recipient's Contact Info.)

RE: (Project's Name and Tracking Number)

Dear (Recipients Name),

   All building and material components will be provided with warranties that you should file upon project close-out.   In order to confirm that the work, as installed, meets the manufacturers' recommendations, certain sections of the technical specifications of the project manual may require inspection by a manufacturer's representative. A written confirmation by the manufacturer's representative of the application and installation of its product in an assembly, or part thereof, provides additional validation of its final performance.

   Certain warranties require a manufacturer's representative site inspection and approval endorsing such application. We will act upon it on your behalf to ensure compliance of all required inspections for this job. I hope this provides you with the necessary information responding to your concern regarding the warranties. Please feel free to call me if you have any additional questions.

Very truly yours,

(Name)
(Title)

Company Name, Address, Tel., Fax., E-mail, Web-Site Address

LETTER TYPE: ADVISORY
ADDRESSED TO: LAWYER
RE: BUILDERS RISK INSURANCE-1

SCENARIO:

An accidental fire devastated a contractor's project. The client places all liability on the contractor, and attempts to recover damages from the contractor's insurance company. The contractor asks for legal advice on this matter.

> **REFERENCE NOTE:**
> Use-BNi-W Form 231, Claim for Damages, if it is required that a contractor recuperate resource and material losses.

COMPANY LETTERHEAD

(Recipient's Name)   (Date)
(Recipient's Title)
(Recipient's Contact Info.)

RE: (Project's Name and Tracking Number)

Dear (Recipient's Name),

Thank you for taking my call regarding the total destruction of the above referenced building project. I understand you are very busy so I will try to summarize the situation in the most efficient manner.

The contract states that it is the owner's responsibility to supply risk insurance in the event of major property damage or loss. Unfortunately, the structure we were contracted to build was completely destroyed by fire. The owner wants to pressure our insurance carrier to accept the claim.

However, our insurance policy will not cover the claim since the fire was not a result of our performance or negligence. The client has informed me that if the insurance does not pay for the damages, and then he will take legal action. I am very concerned and need immediate legal advice. I really would appreciate it if you could respond to this letter at your earliest convenience.

Respectfully,

(name)
(Title)

Company Name, Address, Tel., Fax., E-mail, Web-Site Address

LETTER TYPE: **LEGAL-THREAT RESPONSE**
ADDRESSED TO: **CLIENT**
RE: **BUILDERS RISK INSURANCE-2**

SCENARIO:

The general contractor in the previous letter reviews the contract with his attorney and gets legal advice. After his position has been established, he decides to inform the client.

REFERENCE NOTE:
Use-BNi-W Form 231, Claim for Damages, if it is required that a contractor recuperate resource and material losses.

COMPANY LETTERHEAD

(Recipient's Name)   (Date)
(Recipient's Title)
(Recipient's Contact Info.)

RE: (Project's Name and Tracking Number)

Dear (Recipient's Name),

   I understand your concern regarding the damages that resulted from the fire. However, after careful review of the contract documents and after consulting with counsel, I have determined that it is the owner's responsibility to supply risk insurance.

   You as the owner are required to process a claim with the risk carrier and use the proceeds of the policy to pay me for the cost of rebuilding. If the proceeds of the insurance carrier are insufficient to pay this cost, then by law you take the loss.

   Unfortunately, it appears that you did not purchase the required insurance. At this point, we are ready to respond to your concerns through legal representation. Please let us know how you would like to proceed.

Very truly yours,

(Name)
(Title)

Company Name, Address, Tel., Fax., E-mail, Web-Site Address

LETTER TYPE: **ADVISORY**
ADDRESSED TO: **ATTORNEY**
RE: **WORK DELAYED**

SCENARIO:

A contractor was forced to temporarily stop a grading job due to heavy rain. Now that he is ready to commence the job, the client refuses to pay. The contractor seeks legal advice.

REFERENCE NOTE:
Use-BNi-W Form 322 if using a Notice of Request for Extension of Time.
OR
Use-BNi-W Form 204 if using a Notice of Disclaimer and Protest.
OR
Use-BNi-W Form 200 if you plan to provide a Notification of Data Needed for Service of 20-Day Preliminary Notice.
OR
Use-BNi-W Form 105C for a Legal Perquisite to Filing Claim of Lien and/or Stop Notice-20-Day Preliminary Notice.

COMPANY LETTERHEAD

(Recipient's Name)   (Date)
(Recipient's Title)
(Recipient's Contact Info.)

RE: (Project's Name and Tracking Number)

Dear (Recipient's Name),

Please provide legal advice on the following situation:

We are the contractor grading the above site for a residential project. Unfortunately, due to a prolonged period of heavy rain, we were forced to stop all work. Now that we are ready to re-commence the job and charge for performed services, the client is refusing payment.

The client is financing through a construction loan and explained that the "construction draw" is not allowed at this time since the value of the lot is less than the value of the loan accumulated so far. It is clear that my client has a problem, but I need to pay my subcontractors.

I will follow-up on this message by a phone call. Thank you for your time.

Sincerely,

(Name)
(Title)

Company Name, Address, Tel., Fax., E-mail, Web-Site Address

LETTER TYPE: **LEGAL-PAYMENT REQUEST**
ADDRESSED TO: **CLIENT**
RE: **REFUSED PAYMENT**

SCENARIO:

The general contractor in the previous letter reviews the contract with his attorney and gets legal advice. After his position has been established, he decides to inform the client.

REFERENCE NOTE:
Use-BNi-W Form 322 if using a Notice of Request for Extension of Time.
OR
Use-BNi-W Form 204 if using a Notice of Disclaimer and Protest.
OR
Use-BNi-W Form 200 if you plan to provide a Notification of Data Needed for Service of 20-Day Preliminary Notice.
OR
Use-BNi-W Form 105C for a Legal Perquisite to Filing Claim of Lien and/or Stop Notice-20-Day Preliminary Notice.

COMPANY LETTERHEAD

(Recipient's Name)   (Date)
(Recipient's Title)
(Recipient's Contact Info.)

RE: (Project's Name and Tracking Number)

Dear (Recipient's Name),

After consulting with my lawyer, I would like to clarify my current position on the situation at hand regarding payment refusal. I understand that in order to achieve "construction draws," the value of the lot or building has to equal or exceed the value of the loan proceeds at that particular time. You have stated that since the grading has not been completed, you can't justify value for the draw. However, our contract has a clause that allows for delays caused by inclement weather and events beyond the reasonable control of the contractor.

By virtue of "breach of contract," we have the legal right to stop work and attempt payment collection with accrued interest. Due to the circumstances, we are allowing you to cover the owed amount by (date). Please understand that in order to keep serving you, my workers need to be paid. Please let me know if you have any questions regarding our position.

Yours truly,

(Name)
(Title)

Company Name, Address, Tel., Fax., E-mail, Web-Site Address

LETTER TYPE: **LEGAL DISCLOSURE**
ADDRESSED TO: **CLIENT**
RE: **LEGAL STATEMENT (UNLICENSED -WORK)**

SCENARIO:

After offering an estimate on a residential remodeling job, the contractor realized that the owner may be seeking an unlicensed individual. The contractor decides to politely send the client a legal disclosure statement that may convince the client about hiring unlicensed contractors.

REFERENCE NOTE:
Use-BNi-W Form 272 for a Request for California Contractor's Information

---

COMPANY LETTERHEAD

(Recipient's Name)   (Date)
(Recipient's Title)
(Recipient's Contact Info.)

RE: (Project's Name and Tracking Number)

Dear (Recipient's Name),

Please review the following disclosure regarding the ability to perform work on a residential property. The release of this statement is required by state law.

"STATE LAW REQUIRES ANYONE WHO CONTRACTS TO DO CONSTRUCTION WORK TO BE LICENSED BY THE CONTRACTORS' STATE LICENSE BOARD IN THE LICENSE CATEGORY IN WHICH THE CONTRACTOR IF GOING TO BE WORKING-IF THE TOTAL PRICE OF THE JOB IS $500 OR MORE (INCLUDING LABOR AND MATERIALS). LICENSED CONTRACTORS ARE REQUIRED BY LAWS DESIGNED TO PROTECT THE PUBLIC. IF YOU CONTRACT SOMEONE WHO DOES NOT HAVE A LICENSE, THE CONTRACTORS' STATE LICENSE BOARD MAY BE UNABLE TO ASSIST YOU WITH A COMPLAINT. YOUR ONLY REMEDY AGAINST AN UNLICENSED CONTRACTOR MAY BE IN CIVIL COURT, AND YOU MAY BE LIABLE FOR DAMAGES ARISING OUT OF ANY INJURIES TO THE CONTRACTOR OR HIS OR HER EMPLOYEES. YOU MAY CONTACT THE CONTRACTORS' STATE LICENSE BOARD TO FIND OUT IF THIS CONTRACTOR HAS A VALID LICENSE. THE BOARD HAS COMPLETE INFORMATION ON THE HISTORY OF LICENSED CONTRACTORS, INCLUDING ANY POSSIBLE SUSPENSIONS, REVOCATIONS, JUDGMENTS, AND CITATIONS. THE BOARD HAS OFFICES THROUGHOUT CALIFORNIA. PLEASE CHECK THE GOVERNMENT PAGES OF THE WHITE PAGES FOR THE OFFICE NEAREST YOU OR CALL 1-800-xxx-xxxx FOR MORE INFORMATION."

We hope this helps you in the selection of a contractor. Please don't hesitate to contact me if you have any questions. Thank you.

Sincerely,

(Name)
(Title)

Company Name, Address, Tel., Fax., E-mail, Web-Site Address

LETTER TYPE: **EXPLANATION**
ADDRESSED TO: **CLIENT**
RE: **CLIENT AND PROJECT PROTECTION
BY LAW**

SCENARIO:

The client in the previous letter awarded the licensed contractor the job. However, the client has requested a response explaining what level of protection is involved when hiring a licensed contractor.

REFERENCE NOTE:
Use-BNi-W Form 272 for a Request for California Contractor's Information

---

COMPANY LETTERHEAD

(Recipient's Name)   (Date)
(Recipient's Title)
(Recipient's Contact Info.)

RE: (Project's Name and Tracking Number)

Dear (Recipient's Name),

Thank you for selecting us as the contractor for your remodel. We are confident that you will find that we will protect your interests. We are not only licensed but bonded as well. We are protected by both performance and payment bonds. The performance bond is designed to protect you by insuring project completion. The payment bond protects our subcontractors and material suppliers from not being paid.

We are also insured in order to protect our employees from any accident resulting from the work performed. This will remove any workers' liability from you. The only type of protection our insurance will not provide is loss of property by catastrophic events such an earthquake or fire. We recommend that you buy this protection from your current insurance provider. You could purchase this level of coverage for the extent of time we will work on your job. I hope this information answers your questions.

Once again, thank you for selecting us to serve your current construction needs. We look forward to our pre-construction meeting. If you have any additional questions, please don't hesitate to contact me.

Respectfully,

(Name)
(Title)

Company Name, Address, Tel., Fax., E-mail, Web-Site Address

LETTER TYPE: **PROBLEM INSPECTION RESPONSE**
ADDRESSED TO: **CLIENT**
RE: **MOLD PROBLEM-1**

SCENARIO:

After 9 months of occupancy, the owner of a new residence takes a one week vacation. Upon her return, she finds her home flooded; signs of mold are becoming visible on walls and ceilings. She immediately contacts her contractor. The contractor performs an immediate inspection and responds on his findings:

REFERENCE NOTE:
Use-BNi-W Form 218 for a 48-Hour Notice to Correct Deficiency in Workmanship.
                              OR
Use-BNi-W Form 222 for a Notice of Release of Responsibility for work to be or that has been performed.

COMPANY LETTERHEAD

(Recipient's Name)   (Date)
(Recipient's Title)
(Recipient's Contact Info.)

RE: (Project's Name and Tracking Number)

Dear (Recipient's Name),

Thank you for bringing your mold problem to my attention. After performing an immediate inspection upon notice, we discovered that the cause of the mold problem is due to the fact that your restroom flooded. Unfortunately, the problem was not caused by any construction deficiencies or errors. Furthermore, we can't legally perform abatement on the mold that formed on your walls and ceilings due to special license and insurance provisions. However, as a courtesy to you, we assisted by drying the affected areas. Please review our contract "mold exclusion" clause:

"Cost of correcting/testing/remediation mold/fungus/mildew and organic pathogens unless caused by the sole and active negligence of contractor as a direct result of a construction defect that caused sudden and significant amounts of water infiltration into a part of the structure."

If you had notified us immediately after this incident, we would have responded differently. A response to a sudden leak or flood begins by immediately canceling the source and removing/drying all standing water. Next, any porous material affected such as insulation and drywall needs to be removed. I understand this happened when you were out of town and you notified us within a week from the occurrence. I recommend that you contact your insurance provider. We have documented the situation and would be more than happy to assist you in the selection of a specialized mold-abatement contractor. Please feel free to contact me if you have any questions.

Sincerely,

(Name)
(Title)

Company Name, Address, Tel., Fax., E-mail, Web-Site Address

LETTER TYPE: **WARRANTY DESCRIPTION**
ADDRESSED TO: **CLIENT**
RE: **MOLD PROBLEM-2**

SCENARIO:

The client in the previous "mold problem" letter attempts to use the contractor's warranty to pay for the mold-abatement expense.

REFERENCE NOTE:
Use-BNi-W Form 218 for a 48-Hour Notice to Correct Deficiency in Workmanship.
OR
Use-BNi-W Form 222 for a Notice of Release of Responsibility for work to be or that has been performed.

COMPANY LETTERHEAD

(Recipient's Name)   (Date)
(Recipient's Title)
(Recipient's Contact Info.)

RE: (Project's Name and Tracking Number)

Dear (Recipient's Name),

   Our policy clearly describes the mold exclusions as "problems caused by lack of owner maintenance, owner abuse, owner misuse; damages resulting from mold, fungus and other organic pathogens unless caused by the sole and active negligence of the contractor."

   Our warranties are strictly set by the insurance provider under current applicable coverage policies. That is the reason our contract provides the "mold exclusion" clause. If you wish, I could provide you with my agent's contact information if you have any additional questions.

Respectfully,

(Name)
(Title)

Company Name, Address, Tel., Fax., E-mail, Web-Site Address

LETTER TYPE: **NOTICE**
ADDRESSED TO: **CLIENT**
RE: **VACATING JOBSITE**

SCENARIO:

The contractor writes the following notice to his client as a reminder of a pre-established vacating agreement:

REFERENCE NOTE:
Use-BNi-W Form 214 if you intend to use a Safety Violation Citation.
                              OR
Use-BNi-W Form 222 for a Notice of Release of Responsibility for work to be or that has been performed.

COMPANY LETTERHEAD

(Recipient's Name)   (Date)
(Recipient's Title)
(Recipient's Contact Info.)

RE: (Project's Name and Tracking Number)

Dear (Recipient's Name),

    This is a reminder regarding your temporary move. Per our contract, you have agreed to vacate the house during the project. Please be sure to remove all personal property. Your house shall be vacant and free of all personal property one day prior to our project commencement.

    We cannot be responsible for the protection of your property from dust or construction debris. We apologize for any inconvenience. Please contact us if you have any difficulty with this request.

    Thank you very much for your support and understanding.

Sincerely,

(Name)
(Title)

Company Name, Address, Tel., Fax., E-mail, Web-Site Address

LETTER TYPE:	AGREEMENT SIGNATURE REQUEST
ADDRESSED TO:	CLIENT
RE:	CONFLICT OF DOCUMENTS AGREEMENT

SCENARIO:

A contractor is bidding on a small commercial project. The client gives instructions to change door types. The contractor does not agree with this decision and needs to note it. He writes the following letter to obtain the client's written approval.

> **REFERENCE NOTE:**
> *Use-BNi-W Form 312 as a Notice of Request for Change*
> *In Specifications and Substitutions.*
> *OR*
> *Use-BNi-W Form 315 for Substitution Agreements between the general contractor and subcontractor*

COMPANY LETTERHEAD

(Recipient's Name)   (Date)
(Recipient's Title)
(Recipient's Contact Info.)

RE: (Project's Name and Tracking Number)

Dear (Recipient's Name),

The plans and specifications for your project calls for 45 minute doors. Per our conversation on (date), you directed us to change our bid to reflect non-rated doors. By this order, we are now bound to bid the project with a "Conflict of Documents Agreement." Prior to this decision, we would suggest that you review this change with your architect.

We are not responsible for determining if the owner's changes or suggestions conform to all applicable building codes as we are not the architect. Please confirm in writing that this is your final decision so that we may bid this project according to your request.

Respectfully,

(Name)
(Title)

Company Name, Address, Tel., Fax., E-mail, Web-Site Address

LETTER TYPE: **PAYMENT REQUEST**
ADDRESSED TO: **CLIENT**
RE: **BUILDING PERMIT REIMBURSEMENT**

SCENARIO:

A client refuses to pay for the building permit fees associated with his remodeling project. The contractor asks for the money

> **REFERENCE NOTE:**
> *Use-BNi-W Form 328 for an Application for Payment when requesting payments that need to be recorded.*
>   *OR*
> *Use-BNi-W Form 315 for Substitution Agreements between the general contractor and subcontractor*

COMPANY LETTERHEAD

(Recipient's Name)   (Date)
(Recipient's Title)
(Recipient's Contact Info.)

RE: (Project's Name and Tracking Number)

Dear (Recipient's Name),

It is my understanding that you are not willing to pay for the construction permit fees. I am certain that this may be due to a contract misunderstanding on your part.

As a matter of clarification, per our agreement, the contractor will assist the owner in obtaining the building permit. However, the owner will pay the cost of all governmental permit fees and public/private utility connection fees.

Please review our contract. Your assistance with this payment is critical in maintaining the target schedule. Please call me at your earliest convenience. Thank you.

Respectfully yours,

(Name)
(Title)

Company Name, Address, Tel., Fax., E-mail, Web-Site Address

LETTER TYPE: **NOTICE**
ADDRESSED TO: **CLIENT**
RE: **RESERVING IMPACT COSTS**

SCENARIO:

The following letter is a notice regarding a change order that could produce a ripple effect of additional costs and expenses. The contractor needs to protect the right to make a claim for additional compensation. Without this letter of reservation, the executed change order will prohibit any claim for additional compensation.

REFERENCE NOTE:
Use-BNi-W Form 211 as a Change Order Form for use in the performance of the contract.
OR
Use-BNi-W Form 232 as a Change Order Form if the client requires a quotation for a change in scope of work.
OR
Use-BNi-W Form 226 if you need to give official Notice of a possible cost increase confirmation.
OR
Use-BNi-W Form 230 if you need to give a Notice of Excusable Delay and request for Extension Time

---

COMPANY LETTERHEAD

(Recipient's Name)   (Date)
(Recipient's Title)
(Recipient's Contact Info.)

RE: (Project's Name and Tracking Number)

Dear (Recipient's Name),

We are in receipt of your change order (number) which compensates us for the direct costs of the extra work in that change order. However, as you know, the directive to perform this extra work and the performance of this extra work had such an impact and effect on the balance of the work on this project that we incurred additional costs and expenses as a direct result.

We understand that this change order has been issued so that you can pay us for the direct costs we incurred. However we are executing and returning to you the change order with the understanding that it is limited to that change order alone.

We have also reserved and preserved our claim for the additional costs which we have incurred, and that our right to proceed on our claim for impact costs, attendant field costs, general administration costs, and for an extension of time.

We appreciate your prompt response and cooperation on this matter. Thank you.

Very truly yours,

(Name)
(Title)

Company Name, Address, Tel., Fax., E-mail, Web-Site Address

Letters to Clients

LETTER TYPE: **NOTICE OF PROTEST**
ADDRESSED TO: **CLIENT**
RE: **REDUCTION OF FUNDS**

SCENARIO:

A contractor received a check for an amount less than the amount he was entitled to. This letter gives the necessary notice to protect the right to make a claim for the deducted amount.

REFERENCE NOTE:
Use-BNi-W Form 325 as a Notice of Protest is reduction of funds from a pay estimate or check resulted.
OR
Use-BNi-W Form 282 if using a Notice of Express Trust Fund Delinquent Payment.

COMPANY LETTERHEAD

(Recipient's Name)   (Date)
(Recipient's Title)
(Recipient's Contact Info.)

RE: (Project's Name and Tracking Number)

Dear (Recipient's Name),

This will acknowledge receipt of check number (number) in the amount of ($amount) and we note that you have withheld from us sums which we are entitled to receive for the work we have performed.

These sums are wrongfully withheld from us under the assessment of liquidated damages and we protest such wrongful withholding. We have cashed your check with the full understanding that we have not in any way waived our claim on the withheld funds and we hereby renew our protest and our right to be paid in full. Please let us hear from you at your earliest convenience.

Respectfully,

(Name)
(Title)

Company Name, Address, Tel., Fax., E-mail, Web-Site Address

LETTER TYPE: **AGREEMENT**
ADDRESSED TO: **CLIENT**
RE: **RECEIPT OF FULL PAYMENT**

SCENARIO:

The contractor of the previous letter settled the reduction of funds dispute with the client. The contractor needs to provide the following letter to acknowledge the termination of the dispute.

REFERENCE NOTE:
Use-BNi-W Form 233 as the Agreement for Receipt of Full Payment and Full Release.

OR

Use-BNi-W Form 247 for a Mutual Release form after a dispute settlement.

COMPANY LETTERHEAD

(Recipient's Name)   (Date)
(Recipient's Title)
(Recipient's Contact Info.)

RE: (Project's Name and Tracking Number)

Dear (Recipient's Name),

This letter provides you with a final release of claims arising from the reduction of funds settlement agreement executed on (date). I, (contractor's name), have been paid in full for all labor, subcontract work, equipment, and materials supplied to the above contract and I do hereby release all mechanic's liens, stop notices, equitable lien and labor and material bond rights on the project.

This release is for the benefit of the owner, the prime contractor, the construction lender, and the principal and surety on any labor and material bond posted for the project.

It is understood and agreed that this is a compromise settlement of a disputed claim, and that the payment of the consideration for this release shall not be deemed or construed as an admission of any liability or fact by my signature on this release.

Please don't hesitate to contact me if you should have any questions or concerns. Thank you.

Sincerely,

(Name)
(Title)

Company Name, Address, Tel., Fax., E-mail, Web-Site Address

LETTER TYPE: NOTICE
ADDRESSED TO: CLIENT
RE: VISITOR'S RELEASE

SCENARIO:

The contractor of a new project has been receiving, on the jobsite, vendors and other unscheduled visitors sent by the owner. The contractor is concerned about the liability and decides to take action by informing the owner.

REFERENCE NOTE:
Use-BNi-W Form 214 if you need to issue a Safety Violation Citation to protect you in both liability for injuries and OSHA. requirements.
OR
Use-BNi-W Form 330 for a Visitor's Release form for anyone entering the jobsite area.

COMPANY LETTERHEAD

(Recipient's Name)   (Date)
(Recipient's Title)
(Recipient's Contact Info.)

RE: (Project's Name and Tracking Number)

Dear (Recipient's Name),

Please be aware that visitors entering the jobsite area need to be granted permission in advance by signing a "Visitor's Release" form. The form is designed to protect us from litigation. The form grants permission to the undersigned to enter the premises on the condition of releasing and forever discharging our company, its agents, and all other persons, firms and corporations connected from any form of liability. This will be an enforced condition for all visitors and vendors.

Please keep in mind we do this to promote safety and to protect your interests as well as ours. Please feel free to contact me if you have any specific question regarding this decision. The forms will be made available to you upon request. Thank you for your cooperation.

Very truly yours,

(Name)
(Title)

Company Name, Address, Tel., Fax., E-mail, Web-Site Address

LETTER TYPE: **NOTICE**
ADDRESSED TO: **CLIENT**
RE: **RIGHT OF RESCISSION**

SCENARIO:

A client states that he may want to cancel the project. The contractor has the obligation to inform the client of his rights.

REFERENCE NOTE:
Use-BNi-W Form 279 if you need to deliver the client a Notice of Right of Rescission.

COMPANY LETTERHEAD

(Recipient's Name)   (Date)
(Recipient's Title)
(Recipient's Contact Info.)

RE: (Project's Name and Tracking Number)

Dear (Recipient's Name),

    In response to the concern you expressed regarding a possible cancellation, please note that you have the legal "right of rescission". This gives you the right to rescind the contract before we commence work. As always, any legal document, contract, or state-specific reference should be reviewed by your attorney.

    In any event, please keep in mind that we are here to help you with your construction needs and any other matter in which we could be helpful. Please don't hesitate to contact me if you need my assistance.

Sincerely,

(Name)
(Title)

Company Name, Address, Tel., Fax., E-mail, Web-Site Address

LETTER TYPE: **NOTICE**
ADDRESSED TO: **CLIENT**
RE: **MECHANICS' LIEN LAW-1**

SCENARIO:

Contractor gives a notice to the owner regarding the mechanics' lien law.

REFERENCE NOTE:
Use-BNi-W Form 104-A as the Notice To Owner Regarding Mechanics' Lien Law.

COMPANY LETTERHEAD

(Recipient's Name)   (Date)
(Recipient's Title)
(Recipient's Contact Info.)

RE: (Project's Name and Tracking Number)

Dear (Recipient's Name),

Under the (state, location) Mechanics' Lien Law, any contractor, subcontractor, laborer, supplier, or other person or entity that helps to improve your property, but is not paid for his or her work or supplies, has a right to place a lien on your home, land, or property where the work was performed and to sue you in court to obtain payment.

This means that after a court hearing, your home, land, and property could be sold by a court officer and the proceeds of the sale used to satisfy what you owe. This can happen even if you have paid your contractor in full if the contractor's subcontractors, laborers, or suppliers remain unpaid.

To preserve their rights to file a claim or lien against your property, certain claimants such as subcontractors or material suppliers are each required to provide you with a document called a "preliminary notice." Contractors and laborers who contract with owners directly do not have to provide such notice, since you, as the owner, are aware of their existence. A preliminary notice is not a lien against your property. Its purpose is to notify you of persons or entities that may have a right to file a lien against your property if they are not paid. In order to exercise their lien rights, a contractor, subcontractor, supplier, or laborer must file a mechanics' lien with the county recorder which then becomes a recorded lien against your property. Generally, the maximum time allowed for filing a mechanics' lien against your property is (number) days after substantial completion of your project.

We understand this is a very important matter and would like to assist you in the best possible way. Please let me hear from you if you have any additional questions.

Very truly yours,

Company Name, Address, Tel., Fax., E-mail, Web-Site Address

LETTER TYPE: **NOTICE**
ADDRESSED TO: **CLIENT**
RE: **MECHANICS' LIEN LAW-2**

SCENARIO:

The client of the previous letter has requests that the contractor offer some advice as what to do to avoid the mechanics' lien against his project. The contractor responds.

REFERENCE NOTE:
Use-BNi-W Form 104-A as the Notice To Owner Regarding Mechanics' Lien Law.

COMPANY LETTERHEAD

(Recipient's Name)   (Date)
(Recipient's Title)
(Recipient's Contact Info.)

RE: (Project's Name and Tracking Number)

Dear (Recipient's Name),

   TO INSURE EXTRA PROTECTION FOR YOURSELF AND YOUR PROPERTY, YOU MAY WISH TO TAKE ONE OR MORE OF THE FOLLOWING STEPS:

   1.   Require that your contractor supply you with a payment and performance bond (not a license bond), which provides that the bonding company will either complete the project or pay damages up to the amount of the bond. This payment and performance bond as well as a copy of the construction contract, should be filed with the county recorder for your further protection. The payment and performance bond will usually cost from 1 to 5 percent of the contract amount depending on the contractor's bonding ability. If a contractor cannot obtain such bonding, it may indicate his or her financial incapacity.

   2.   Require that payments be made directly to subcontractors and material suppliers through a joint control. Funding services may be available for a fee in your area, which will establish voucher or other means of payment to your contractor. These services may also provide you with lien waivers and other forms of protection. Any joint control agreement should include the addendum approved by the registrar.

   3.   Issue joint checks for payment, made out to both your contractor and subcontractors or material suppliers involved in the project. The joint checks should be made payable to the persons or entities which send preliminary notices to you. Those persons or entities have indicated that they may have lien rights on your property; therefore you need to protect yourself. This will help to insure that all persons due payment are actually paid.

   I hope you find this advice useful. Don't hesitate to contact me if I can be of further assistance.

Respectfully,

(Name)
(Title)

Company Name, Address, Tel., Fax., E-mail, Web-Site Address

LETTER TYPE: **NOTICE**
ADDRESSED TO: **CLIENT**
RE: **MECHANICS' LIEN LAW-3**

SCENARIO:

The client in the previous letter has requested that the contractor offer some advice as what to do to avoid a mechanics' lien against his project. The contractor responds.

> *REFERENCE NOTE:*
> *Use-BNi-W Form 104-A as the Notice To Owner Regarding Mechanics' Lien Law.*

COMPANY LETTERHEAD

(Recipient's Name)   (Date)
(Recipient's Title)
(Recipient's Contact Info.)

RE: (Project's Name and Tracking Number)

Dear (Recipient's Name),

   I will be more than happy to describe the waivers and liability releases. To protect yourself under this option, you must be certain that all material suppliers, subcontractors, and laborers have signed the waiver and release form. If a mechanics' lien has been filed against your property, it can only be voluntarily released by a recorded release of mechanics' lien signed by the person or entity that filed the mechanic's lien against your property unless the lawsuit to enforce the lien was not filed in a timely manner. You should not make any final payments until any and all such liens are removed. You should consult an attorney if a lien is filed against your property.

   I hope this answered your question. Please feel free to contact me if you should have any further questions or concerns.

Very truly yours,

(Name)
(Title)

Company Name, Address, Tel., Fax., E-mail, Web-Site Address

LETTER TYPE: NOTICE
ADDRESSED TO: CLIENT
RE: STOP NOTICE (AGAINST CONSTRUCTION FUNDS)

SCENARIO:

A general contractor is exercising his right to send the client a stop notice for funds retained against completed work on a project. He writes the following cover letter.

REFERENCE NOTE:
Use-BNi-W Form 107 as the Stop Notice Form (Legal Notice to Withhold Construction Funds on Public or Private Work)
OR
Use-BNi-W Form 108 as the Stop Notice Bond (Legal Notice to Withhold Construction Funds)

COMPANY LETTERHEAD

(Recipient's Name)   (Date)
(Recipient's Title)
(Recipient's Contact Info.)

RE: (Project's Name and Tracking Number)

Dear (Recipient's Name),

Enclosed please find the stop notice for the above referenced project. Be aware that at a copy of this notice has been sent to your architect. This stop notice is against the construction funds and will be served upon the construction lender holding the funds. This notice is bonded and accompanied by a bond with a penal sum equal to one and one-quarter (1 1/4) times the amount of the claim noted on the form.

Please let us hear from you regarding this situation so that we may clear the issues and continue serving you. Thank you for your cooperation.

Very truly yours,

(Name)
(Title)

Company Name, Address, Tel., Fax., E-mail, Web-Site Address

LETTER TYPE: **ADVISORY**
ADDRESSED TO: **CLIENT**
RE: **BUILDING WITHOUT PERMIT**

SCENARIO:

The contractor of choice for a building addition project has discovered that the building has recently been added to without a building permit or inspections. He is concerned for the client and will not do anything until the client resolves this situation with the building department. He offers some advice.

COMPANY LETTERHEAD

(Recipient's Name)   (Date)
(Recipient's Title)
(Recipient's Contact Info.)

RE: (Project's Name and Tracking Number)

Dear (Recipient's Name),

    Thank you for selecting us as the contractor for your project. At this point we need to bring to your attention some unfortunate news. We have discovered that the previous addition was built without a permit. Due to its recent completion, we will face confrontation with the city officials and inspectors when disclosing the existing conditions of your project. Unfortunately this may even prevent us from working on it. Furthermore, you are liable and could face fines for building without permits and inspections. When the building official discovers that the previous addition was done without a permit, he may require disassembling and reconstructing the project with proper inspection records.

    At this point we suggest that you contact the building department and attempt some level of negotiation with them before anything else is done. Please remember that we are here to assist you in whatever way we can. Feel free to call me if you have any questions.

Very truly yours,

(Name)
(Title)

Company Name, Address, Tel., Fax., E-mail, Web-Site Address

LETTER TYPE: AWARD RESPONSE
ADDRESSED TO: CLIENT
RE: CONTRACT FUND REBATE

SCENARIO:

A general contractor was selected to build a commercial development project. However, the owner conditions this offer by imposing a rebate on the contract price. The contractor wants the project but has to refuse any type of rebate as forbidden and penalized by law. The contractor informs the client:

COMPANY LETTERHEAD

(Recipient's Name)   (Date)
(Recipient's Title)
(Recipient's Contact Info.)

RE: (Project's Name and Tracking Number)

Dear (Recipient's Name),

Thank you very much for the opportunity to be the contractor for your new project. We realize it is a great opportunity, however we have to inform you that the (state) penal code states the following:

"Any person who receives money for the purpose of obtaining or paying for services, labor, materials or equipment incident to constructing improvements on real property and willfully rebates any part of the money to or on behalf of anyone contracting with such person, for provision of the services, labor, materials, or equipment for which the money was given, shall be guilty of a misdemeanor; provided, however, that normal trade discount for prompt payment shall not be considered a violation of this section."

Please let me know if we can find any other type of relationship that may benefit both of our companies. Please keep in mind that we are here to serve your construction needs and protect your legal rights as well as ours. Thank you once again for your generous consideration and for selecting us as your contractor. I would like to offer you a construction business plan that would assure you some substantial savings. I look forward to hearing from you.

Very truly yours,

(Name)
(Title)

Company Name, Address, Tel., Fax., E-mail, Web-Site Address

LETTER TYPE: **ADVISORY**
ADDRESSED TO: **CLIENT**
RE: **FINAL BUILDING PRICE**

SCENARIO:

The client of a fast-food restaurant has complained to the contractor regarding the final project cost. The client argues that the architect's estimate was 15% below the final cost. This situation is delicate since the client requested a construction loan for the architect's estimate plus only an additional 5% for contingency. The client wants to process a claim or take legal action to collect the difference. In order to protect his company, the contractor writes the following letter:

REFERENCE NOTE:
Use-BNi-W Form 320 if you need to obtain a Notice of Directive or Communication from the client or design professional.

COMPANY LETTERHEAD

(Recipient's Name)   (Date)
(Recipient's Title)
(Recipient's Contact Info.)

RE: (Project's Name and Tracking Number)

Dear (Recipient's Name),

We are confident that our services to you have been performed with honesty and professionalism. We have submitted a log of change orders to both you and the architect in order to keep an accurate record of any cost increase and allowances. I understand that the total project cost exceeded the architect's estimate by 15%.

In our particular situation, the architect has provided an estimate that by law is only justified as an estimate and not a final and precise quote. The only protection you would have against the cost difference is a "Cost-Plus Contract". You may want to review the contract with the architect. We can only offer you some suggestions, but we recommend that you seek advice from your attorney. Please feel free to contact me if you need to discuss this.

Very truly yours,

(Name)
(Title)

Company Name, Address, Tel., Fax., E-mail, Web-Site Address

LETTER TYPE: **RESPONSE**
ADDRESSED TO: **CLIENT**
RE: **STRUCTURAL DEFICIENCY**

SCENARIO:

A real estate broker and investor invite bidders for a residential property remodel. According to the broker, it should be a simple job with emphasis on aesthetics. After talking to the broker, a contractor decides to visit the site before offering an estimate. At the site, he notices that there are major structural deficiencies involved. These deficiencies are very costly. The broker's lack of understanding caused him to underestimate the scope and budget. The contractor describes the situation to the client.

COMPANY LETTERHEAD

(Recipient's Name)   (Date)
(Recipient's Title)
(Recipient's Contact Info.)

RE: (Project's Name and Tracking Number)

Dear (Recipient's Name),

    We have inspected the property referenced above. Before we make any commitment to this remodel, please be aware that major problems were detected. We documented (number) locations showing evidence of (description of the deficiencies). These observations are indicative of major structural problems that need to be discussed with a structural engineer. It now appears that this project will not be an easy and inexpensive remodel as you previously suggested.

    Please review the existing conditions and determine if we should demolish and start from scratch. This option is extreme, but will surely give you peace of mind that everything is up to code and structurally sound. Please let me know how we can be of further assistance. Thank you for the opportunity to bid this job.

Very truly yours,

(Name)
(Title)

Company Name, Address, Tel., Fax., E-mail, Web-Site Address

LETTER TYPE: NOTICE
ADDRESSED TO: CLIENT
RE: MILDEW PROBLEM

SCENARIO:

While servicing a maintenance routine for a major property management company, the contractor discovers mildew forming on the perimeter walls and ceiling of an assisted living complex. After investigating the problem, the contractor realizes that the problem was caused by inadequate ventilation. The contractor reports this to the client.

REFERENCE NOTE:
Use-BNi-W Form 320 if you need to obtain a Notice of Directive or Communication from the client or design professional.

COMPANY LETTERHEAD

(Recipient's Name)   (Date)
(Recipient's Title)
(Recipient's Contact Info.)

RE: (Project's Name and Tracking Number)

Dear (Recipient's Name),

The reason I am sending you this letter along with the invoice for (month) maintenance repairs is to inform you of a problem in your building.

This month we started with the maintenance routine for the (location and building description) managed by your company. We discovered a consistent problem on (location). We noticed that on those units, the wall paper was peeling off from top to bottom along perimeter walls. The drywall was completely damp. In some cases mold was already present. We took a closer look into the attic and followed the mildew trace where we found that it was caused by inadequate ventilation. We could not find soffit vents anywhere on these units. This is an obvious construction defect.

At this point we suggest that you involve your insurance company and lawyer to determine a course of action. Please let me know if there is anything we can do to assist you.

Very truly yours,

(Name)
(Title)

Company Name, Address, Tel., Fax., E-mail, Web-Site Address

LETTER TYPE: **NOTICE**
ADDRESSED TO: **CLIENT**
RE: **INADEQUATE ELECTRICAL SERVICE**

SCENARIO:

A contractor is approached by a former client with a request for an addition that will house sophisticated appliances and electrical equipment. After reviewing the equipment power requirements, the contractor inspects the existing electrical service and realizes that it will not be sufficient.

COMPANY LETTERHEAD

(Recipient's Name)   (Date)
(Recipient's Title)
(Recipient's Contact Info.)

RE: (Project's Name and Tracking Number)

Dear (Recipient's Name),

Thank you for choosing us as your contractor. We will be more than happy to assist you with the addition.

However, let me explain the challenge we will face.

We understand your appliance and equipment requirements. Unfortunately, your house has only a 120 volt service only. You need to contact the local power company to upgrade your service to a 240-volt service.

The electrical panel on your house is also limited to 60-amps. The power demand you need on your new appliances and equipment will require a 100-amp, circuit breaker service panel. We can certainly help you achieve the new requirements.

The first step will be to contact the local power company and request the service upgrade. At that point we can connect our new panel. Please call me if you need further assistance or if you have any questions or concerns.

Very truly yours,

(Name)
(Title)

Company Name, Address, Tel., Fax., E-mail, Web-Site Address

LETTER TYPE: RESPONSE
ADDRESSED TO: CLIENT
RE: PERMIT

SCENARIO:

A contractor's former client is remodeling the restrooms of his house. He asks the contractor if a permit would be required and explains that new fixtures will be used. The contractor offers some advice.

COMPANY LETTERHEAD

(Recipient's Name)   (Date)
(Recipient's Title)
(Recipient's Contact Info.)

RE: (Project's Name and Tracking Number)

Dear (Recipient's Name),

   I will be more than happy to assist you with this project. Please note that any interior remodeling affecting the structural integrity of the building requires a permit. If the remodel is cosmetic and will not affect the structural integrity, then you are not required to have one. Regarding your concern for a permit on the new fixtures, you don't need a permit as long as the new fixtures will replace existing ones. However, if you increase the number of fixtures then you alter the water demand, resulting in the need for a plumbing permit. When applying for a permit, you will be asked to provide a water meter card showing the total number of existing and new fixtures. You will also be asked to show a plan with dimensions from the wall to the centerline of the fixture and clearances in front of each fixture.

   Changing, adding, or extending pipes requires the use of a permit. For example, each new fixture requires connection to existing vent stacks and in some cases connections to new ones. Please let me know in more detail what your project entails. I hope to be of further help to you.

Very truly yours,

(Name)
(Title)

Company Name, Address, Tel., Fax., E-mail, Web-Site Address

LETTER TYPE: NOTICE AND REQUEST
ADDRESSED TO: CLIENT
RE: ROOFING INSPECTION

SCENARIO:

In the process of a residential remodel and building addition, the contractor discovers significant damage to the existing roof. Even though this part of the residence is not affected by the remodel, the contractor needs to report the findings to the client. This also offers the opportunity to revise the contract to include additional fees for the unexpected and unforeseen conditions.

COMPANY LETTERHEAD

(Recipient's Name)   (Date)
(Recipient's Title)
(Recipient's Contact Info.)

RE: (Project's Name and Tracking Number)

Dear (Recipient's name),

   The reason I am writing you this letter is to inform you that there may be additional work to be performed on your existing roof. As we established in our agreement, the contractor is not liable for existing conditions.

   We recently inspected your roof, and must inform you that there is major deterioration of the shingles, underlayment, and plywood. It is not uncommon to find these types of problems on older houses. At this point, we suggest that you replace the deteriorated roof components to be consistent with your new project.

   We can offer you a competitive price if you choose to select us for this task. We are committed to use the best materials to ensure the best possible results on our jobs.

   Please let me know how you want to proceed. Thank you.

Respectfully,

(Name)
(Title)

Company Name, Address, Tel., Fax., E-mail, Web-Site Address

LETTER TYPE: SUPPORTIVE
ADDRESSED TO: CLIENT
RE: PERFORMANCE-1

SCENARIO:

The owner of construction company "A" met with a former school district client. The former client explains the nightmares he experienced in the construction of a high school project executed through contractor "B". The school district has sued, and contractor "B" has involved the architect and the inspector in the litigation process. Apparently the inspector failed in performing due diligence and acting on behalf of the client.

The owner of company "A" offers his support while using good marketing techniques.

COMPANY LETTERHEAD

(Recipient's Name)   (Date)
(Recipient's Title)
(Recipient's Contact Info.)

RE: (Project's Name and Tracking Number)

Dear (Recipient's name),

We completely agree with you regarding your issues of poor inspection of the installed concrete, this is indicative of the fact that inspection is not an easy job. It requires people who are well trained and properly supported by their employers. It requires properly prepared, realistic construction documents, professionally administered. It requires capable supervision on the part of the contractor.

Proper inspection insures that the owner gets the value and quality is paying for, within the design intent required by the architect and construction industry standards.

We understand your position and hope for a speedy resolution. Please be aware that we are a high quality construction company with a good professional portfolio. Our 25 years of experience has enabled us to establish an excellent reputation with architects and inspectors. Please let me know if there is anything we could do to assist you.

Very truly yours,

(Name)
(Title)

Company Name, Address, Tel., Fax., E-mail, Web-Site Address

LETTER TYPE: **SUPPORTIVE**
ADDRESSED TO: **CLIENT**
RE: **PERFORMANCE-2**

SCENARIO:

The client on the previous letter is very appreciative of contractor's "A" support. However, he asks for some advice in hiring a good inspector. The contractor responds.

COMPANY LETTERHEAD

(Recipient's Name)   (Date)
(Recipient's Title)
(Recipient's Contact Info.)

RE: (Project's Name and Tracking Number)

Dear (Recipient's name),

I will be more than happy to offer you some advice. As you have experienced, the client needs to be aware of the inspector's background and rely on the architect for his input. The construction of a project requires the integrated efforts of a design team composed of competent individuals. The complexity of the disciplines that interact in the construction process necessitates the division of responsibility to different team members based on expertise. Although the project designer conceives the design concept, the actual project realization and delivery requires the technical and management capabilities offered by other team members.

Here are a few of the many ways to employ the construction inspector: The owner engages the construction inspector for the project, and the construction inspector works under the direction of the architect. In some instances this is a legal requirement.

The design professional is often required to select the construction inspector and engages him under a mutually agreed upon extension of the design professional's agreement with the owner. The architect may select and employ the construction inspector for the project, and often the selected person may be a fully qualified member of the design professional's staff. In some cases, the owner maintains a staff of inspectors and assigns one or more to the project.

The inspector prevents problems and avoids misunderstandings by continually reviewing the construction documents and working in conjunction with the superintendents and the subcontractors. The inspector should look ahead and be fully acquainted with the construction documents and all phases of the work, thus avoiding costly mistakes and foreseeing bottlenecks as they occur. I hope this information is useful. Please let me know if there is anything else we can do.

Very truly yours,

(Name)

(Title)

Company Name, Address, Tel., Fax., E-mail, Web-Site Address

LETTER TYPE: **NOTICE**
ADDRESSED TO: **CLIENT**
RE: **PROGRESS PAYMENTS**

SCENARIO:

Per contractual agreement, the general contractor is sending a reminder of the payment arrangement to the client's accounting department.

COMPANY LETTERHEAD

(Recipient's Name)   (Date)
(Recipient's Title)
(Recipient's Contact Info.)

RE: (Project's Name and Tracking Number)

Dear (Recipient's name),

   Please find enclosed copies of all supporting documentation sent by our company that establishes the monthly obligations due for the construction of the above referenced project.

   Please keep in mind that these invoices will be sent the (date) of each month. The amount of the payment will be based on total cost of work at each particular payment period plus the contractor's percentage fee.

   For further billing clarification, a "schedule of values" will be provided with each invoice along with the reimbursable expense items. We appreciate your support and look forward to serving you. Please feel free to contact me if you have any questions.

Very truly yours,

(Name)
(Title)

Company Name, Address, Tel., Fax., E-mail, Web-Site Address

LETTER TYPE: NOTICE
ADDRESSED TO: CLIENT
RE: PROJECT WARRANTIES AND CLOSE-OUT

SCENARIO:

The contractor for a facility is arriving at the completion date. He has decided to inform the client in advance in   order   to   provide   some critical information the client will need to be aware of.

COMPANY LETTERHEAD

(Recipient's Name)   (Date)
(Recipient's Title)
(Recipient's Contact Info.)

RE: (Project's Name and Tracking Number)

Dear (Recipient's name),

Congratulations on your new building! We appreciate the fact that you selected us as your contractor for this project. In order to best serve you and maintain a high level of satisfaction, we look forward to assist you with the close-out process. Shortly, we will send you all manufacturer warranties along with our limited warranty against material defects on all labor and materials supplied by ourselves and the subcontractors. Our warranty will be effective for one year upon substantial completion of your project. Any other technical and operating instructions will be attached to each piece of equipment, and a copy will be provided for your records.

All as-built drawings will be sent back to the architect along with any outstanding paperwork. We will keep a copy in the maintenance room for the purpose of building operations. Please keep in mind that this building is automated and practically all information regarding building operations can be accessed through the systems network. Please contact me to arrange the training session for the systems.

Very truly yours,

(Name)
(Title)

Company Name, Address, Tel., Fax., E-mail, Web-Site Address

LETTER TYPE: **REQUEST**
ADDRESSED TO: **CLIENT**
RE: **SCOPE OF WORK-1**

SCENARIO:

A contractor is approached by a homeowner with a request to provide an estimate on a minor building addition project. The contractor needs more detailed information from the owner or the architect in order to best negotiate the price for this job.

COMPANY LETTERHEAD

(Recipient's Name)   (Date)
(Recipient's Title)
(Recipient's Contact Info.)

RE: (Project's Name and Tracking Number)

Dear (Recipient's name),

   We had the opportunity of meeting with you last week for a pre-bid job walk. At this time we are requesting a full scope of work from you. We understand this is should be a minor building addition.

   However, we would like to provide you with the best possible service by determining what type of agreement would be required based on the scale of this project. From what we know so far, a labor-only agreement would probably offer you a better deal. I look forward to hearing from you. Thank you.

Very truly yours,

(Name)
(Title)

Company Name, Address, Tel., Fax., E-mail, Web-Site Address

LETTER TYPE: **REQUEST**
ADDRESSED TO: **CLIENT**
RE: **SCOPE OF WORK-2**

SCENARIO:

The client on the previous letter provides the contractor with plans for the building addition project. After careful review, the contractor writes the following letter.

..

COMPANY LETTERHEAD

(Recipient's Name)   (Date)
(Recipient's Title)
(Recipient's Contact Info.)

RE: (Project's Name and Tracking Number)

Dear (Recipient's name),

   After careful review of the plans, we have determined that a labor-only agreement would offer you the best possible situation.

   Under this arrangement you would be responsible for furnishing all material and equipment for the job. No materials of any kind would be purchased by me. This would provide you substantial savings. We are also enclosing a labor and materials estimate for your review and comparison.

   We are here to serve you and make this a positive experience. Please let me know what type of agreement works best for you. Thank you for your support.

Very truly yours,

(Name)
(Title)

Company Name, Address, Tel., Fax., E-mail, Web-Site Address

Chapter Two
Letters to Field Professionals

This chapter is devoted to communication between architects, engineers, subcontractors, inspectors and building/planning officials. The tone of these letters is professional and cooperative. The Contractor demonstrates knowledge and experience combined with a team-player attitude to resolve the issue at hand. These letters are extremely effective when used in combination with the forms listed below the *Letter Scenarios*.

LETTER TYPE: **JUSTIFICATION**
ADDRESSED TO: **ARCHITECT**
RE: **CLARIFICATION ON DAMAGED WORK**

SCENARIO:

The architect of a retail project did not approve the concrete slab. The general contractor is convinced that the deficiencies are purely cosmetic and not structural. The contractor responds.

COMPANY LETTERHEAD

(Recipient's Name)   (Date)
(Recipient's Title)
(Recipient's Contact Info.)

RE: (Project's Name and Tracking Number)

Dear (Recipient's Name),

In response to your concern regarding the new concrete slab, I am convinced there is nothing structurally deficient. The cracks and irregularities you noted on your site visit are typical, slightly beyond what would be expected. I understand you have the contractual right to reject work, but I am asking you to reconsider.

I propose that instead we inject an epoxy filler into the larger cracks and grind some of the worst irregularities. We understand your testing engineer will be arriving today to sample the concrete. I will call you later today to discuss the resolution of this problem further.

Thank you.

Best Regards,

(Name)
(Title)

Company Name, Address, Tel., Fax., E-mail, Web-Site Address

LETTER TYPE: **CONFIRMATION REQUEST**
ADDRESSED TO: **ARCHITECT**
RE: **DEMOLITION DRAWINGS**

SCENARIO:

After careful review of the architect's drawings, the general contractor has questions regarding the extent of the demolition scope of work. He writes the architect the following letter.

REFERENCE NOTE:

Use- BNi-W Form 320 as a Notice of Directive or Communication if you need a written record of the architect, engineer or owner's order approval.

OR

Use- BNi-W Form 321 as a Notice of Reserving Impact Costs when additional costs and expenses on the remaining work arise.

OR

Use- BNi-W Form 103 as an Extra Work Order when extra work not covered by the original bid and contract is called for and should be executed.

COMPANY LETTERHEAD

(Recipient's Name)   (Date)
(Recipient's Title)
(Recipient's Contact Info.)

RE: (Project's Name and Tracking Number)

Dear (Recipient's name),

After carefully reviewing your demolition sheets, we are in need of some clarification before starting the job. We have estimated that about 75% of the existing building is to remain unaffected by demolition. This encompasses the existing building structural system composed of concrete columns and lightweight concrete slabs over metal deck. The building envelope, originally built with IFS cladding and stone veneer seem to remain in the north, east, and west elevations. The south elevation shows demo work on the entrance (lobby area), and facade treatments.

The interior demolition scope of work is clearly shown on the drawings. However, we request that you specify the exact facade elements to be removed and the extent of the new openings to be provided in the main entrance area. At this time, we are in need of an urgent coordination meeting. Thank you for your assistance.

Sincerely,

(Name)
(Title)

Company Name, Address, Tel., Fax., E-mail, Web-Site Address

LETTER TYPE: NOTICE
ADDRESSED TO: **ARCHITECT**
RE: **INWARD BUILDING-1**

SCENARIO:

During a commercial remodeling project, the contractor performs a preliminary job survey to assure that the plans and specifications are complete and reflect an accurate scope of work before submitting an estimate for the job. At the site, he observes that one of the load bearing walls is showing an inward bulge. He immediately notifies the architect and owner.

REFERENCE NOTE:

Use- BNi-W Form 320 as a Notice of Directive or Communication if you need a written record of the architect, engineer or owner's order approval.

OR

Use- BNi-W Form 321 as a Notice of Reserving Impact Costs when additional costs and expenses on the remaining work arise.

OR

Use- BNi-W Form 103 as an Extra Work Order when extra work not covered by the original bid and contract is called for and should be executed.

COMPANY LETTERHEAD

(Recipient's Name)   (Date)
(Recipient's Title)
(Recipient's Contact Info.)

RE: (Project's Name and Tracking Number)

Dear (Recipient's name),

During our field survey, we inspected all perimeter load bearing masonry walls. The south wall shows a minor inward bulge that was possibly the result of the expansive soil encountered at the site.

The wall can be repaired at an additional cost. However, proper drainage must be considered in order to avoid this problem from reoccurring. We have asked the architect to inspect the wall and, if needed, consult with a structural engineer to assess the condition. Please advise.

Sincerely,

(Name)
(Title)

Company Name, Address, Tel., Fax., E-mail, Web-Site Address

LETTER TYPE: RESPONSE
ADDRESSED TO: ARCHITECT
RE: INWARD BUILDING-2

SCENARIO:

The contractor in the previous letter obtains a response from the architect. The architect will hire a structural engineer to perform and assess the masonry wall problem. The design strategy and drawings will be provided by the architect through a supplemental addendum to the contract. In the interim, the architect asks the contractor for a repair approach suggestion based on field experience. The contractor responds.

COMPANY LETTERHEAD

(Recipient's Name)   (Date)
(Recipient's Title)
(Recipient's Contact Info.)

RE: (Project's Name and Tracking Number)

Dear (Recipient's name),

I will be pleased to offer you some suggestions regarding your structural issues.

We experienced a very similar situation when remodeling (name project and location). For that project, the bulge in the cmu wall did not exceed two inches. The structural engineer instructed us to shore the wall with galvanized steel beams set into independent concrete pads, equally spaced along the wall. These beams were tightly secured to the cmu wall. We also reinforced the ceiling rafters with wood blocking.

If the bulge exceeds two inches, then other solutions may be necessary. An additional way to stabilize a cmu wall with more severe bulging is to provide a sister wall. Even though the sister wall is a more expensive solution, it offers a new reinforced load bearing wall. Another factor to consider is excavating in order to remove the expansive soil that originally caused the problem. We would use gravel backfill and provide adequate drainage.

Hopefully, this particular condition will not exceed the two inches allowed for shoring. I hope this provides you with some options. I look forward to the engineer's final assessment.

Sincerely,

(Name)
(Title)

Company Name, Address, Tel., Fax., E-mail, Web-Site Address

LETTER TYPE: **APPROVAL REQUEST**
ADDRESSED TO: **ARCHITECT**
RE: **PARTITION REPAIRS**

SCENARIO:

The client of an office tenant improvement has rejected the work done on all interior partitions. The architect and contractor objected by explaining that the cosmetic damages can be easily repaired and will be unnoticeable when completed. The architect verbally commits to this determination in favor of the contractor. Even though the contractor is grateful to the architect, he needs to obtain written approval in order to proceed with the partition repairs.

REFERENCE NOTE:

Use- BNi-W Form 320 as a Notice of Directive or Communication if you need a written record of the architect, engineer or owner's order approval.

OR

Use- BNi-W Form 321 as a Notice of Reserving Impact Costs when additional costs and expenses on the remaining work arise.

OR

Use- BNi-W Form 103 as an Extra Work Order when extra work not covered by the original bid and contract is called for and should be executed.

COMPANY LETTERHEAD

(Recipient's Name)   (Date)
(Recipient's Title)
(Recipient's Contact Info.)

RE: (Project's Name and Tracking Number)

Dear (Recipient's name),

 I consider myself fortunate to work with you as the architect for this project. There is no question that a firm and fair contract administration in the hands of a competent architect is the best asset in situations like the one we are experiencing. We agree that the situation is justified by repairing, and not replacing, the partitions.

 At this time, and based on your determination, we are asking you to convey your decision to the client. We understand the client is insisting that all interior partitions be replaced. However, I strongly believe that after the repair, the existing partitions will perform their intended purpose, demonstrating that we have substantially performed the contract.

 I look forward for the client's written approval.

Sincerely,

(Name)
(Title)

Company Name, Address, Tel., Fax., E-mail, Web-Site Address

LETTER TYPE:	NOTICE
ADDRESSED TO:	ARCHITECT
RE:	UNTREATED WOOD

SCENARIO:

A contractor for a residential remodel reports on the existing untreated wood deterioration. Along with the report, possible solutions that will directly increase the project's cost are discussed.

COMPANY LETTERHEAD

(Recipient's Name)   (Date)
(Recipient's Title)
(Recipient's Contact Info.)

RE: (Project's Name and Tracking Number)

Dear (Recipient's name),

Per our discussion this morning, please be aware that the untreated wood found on the (location) will continue to attract termites. Before we can start this job, we need to abate the termite and mold problem.

I recommend replacing all damaged floor joists with rated wood. Since the grade is no more than 12 inches, we recommend excavating to a minimum depth of 18 inches so as to provide greater separation and avoid changing all subfloor framing. We could also provide additional termite protection by installing a vapor retarder on top of the soil. Where we encounter water puddles, we can install drains.

Please contact me at your earliest convenience to discuss the course of action and cost.

Sincerely,

(Name)
(Title)

Company Name, Address, Tel., Fax., E-mail, Web-Site Address

LETTER TYPE: **RFI**
ADDRESSED TO: **ARCHITECT**
RE: **MISTAKE ON PLANS**

SCENARIO:

A contractor has detected a possible mistake on the structural drawings. He needs to release an RFI to obtain determination from the architect and structural engineer. He writes the following e-mail with its appropriate RFI form attached.

COMPANY LETTERHEAD

(Recipient's Name)   (Date)
(Recipient's Title)
(Recipient's Contact Info.)

RE: (Project's Name and Tracking Number)

Dear (Recipient's name),

    Sheet (number) shows a foundation plan that references to detail (number) on sheet (number). The detail calls for (describe the item in question, i.e. "5/8-inch anchor bolts no more than 6 feet apart"). We feel that this is an error.

    Typically in this seismic zone, a 4 feet spacing is called for. Please review your detail and provide your determination through RFI (number) attached to this e-mail. Thank you.

Sincerely,

(Name)
(Title)

Company Name, Address, Tel., Fax., E-mail, Web-Site Address

LETTER TYPE: **JUSTIFICATION**
ADDRESSED TO: **ARCHITECT**
RE: **ALTERNATE PRODUCT JUSTIFICATION**

SCENARIO:

Upon occupation of a new building, the owner observes that the windows are not compatible with the windows shown on the drawings. He addresses this issue with the architect, and the architect passes the responsibility to the contractor. The contractor responds.

> **REFERENCE NOTE:**
> *Use- BNi-W Form 312 as a Notice of Request for Change in Specifications and Substitutions.*

COMPANY LETTERHEAD

(Recipient's Name)   (Date)
(Recipient's Title)
(Recipient's Contact Info.)

RE: (Project's Name and Tracking Number)

Dear (Recipient's name),

    Thank you for bringing your concern to my attention. When we bid the project, we obtained sub-bids from three material suppliers. The bids ranged from ($cost) to ($cost) per window. We were trying to get the contract so we used the low bidder.

    However, this decision also allowed us to offer you an allowance as a reserve for delays and discrepancies. Please take into consideration that the allowance was utilized in your favor when you wanted to upgrade the bathroom finishes.

    Also, the shop drawings for the final window selection were reviewed and approved by the architect. If you feel uncomfortable with the performance of the product, please inform the manufacturer before the warranty expires.

    I will be sure to get you the manufacturer's contact information. Please call me if you have any additional questions. Thank you.

Sincerely,

(Name)
(Title)

Company Name, Address, Tel., Fax., E-mail, Web-Site Address

LETTER TYPE: CLARIFICATION
ADDRESSED TO: ARCHITECT
RE: FUNDING INCREASE

SCENARIO:

The contractor for a major development project is responding to a cost increase concern from the developer and engineer.

REFERENCE NOTE:
Use- BNi-W Form 320 as a Notice of Directive or Communication if you need a written record of the architect, engineer or owner's order approval.
OR
Use- BNi-W Form 321 as a Notice of Reserving Impact Costs when additional costs and expenses on the remaining work arise.
OR
Use- BNi-W Form 103 as an Extra Work Order when extra work not covered by the original bid and contract is called for and should be executed.

COMPANY LETTERHEAD

(Recipient's Name)   (Date)
(Recipient's Title)
(Recipient's Contact Info.)

RE: (Project's Name and Tracking Number)

Dear (Recipient's Name),

Now that we have explained this issue to you and the developer, we need to commit as a team to a revision of costs, budget reviews, and revised construction timetables.

The proposal we sent you regarding the gas line project requires at least three "spurs" on the main pipeline assuring that the pipeline could be expanded to provide for future expansion.

This pipeline faces some challenges that we are confident we can resolve with adequate funding. In addition, be confident that you are working with one of the most respected sources in the public works field. Please review the proposal's extra costs and let me know if the shortage of funding can be resolved. Thank you for your assistance.

Respectfully,

(Name)
(Title)

Company Name, Address, Tel., Fax., E-mail, Web-Site Address

LETTER TYPE: **CLARIFICATION**
ADDRESSED TO: **ARCHITECT**
RE: **DISCOUNT DENIAL**

SCENARIO:

The client of a major demolition project attempts to get a discount for the potential material recycling. The contractor responds.

COMPANY LETTERHEAD

(Recipient's Name)   (Date)
(Recipient's Title)
(Recipient's Contact Info.)

RE: (Project's Name and Tracking Number)

Dear (Recipient's Name),

   In response to your observations on field cost saving practices, let me explain some of the realities we face in the discharge of materials.

   Contrary to your opinion, we still have to pay out of pocket expenses of ($amount) per load in average to crush and dispose of concrete debris. Please understand that material disposal is expensive, and we want to achieve it in the fewest trips. However, we will make every effort to recycle the steel components, and this may help off-set the disposal cost.

   Ordinarily, this does not contribute in any way towards a significant profit. We calculate our labor and disposal costs and subtract the recycling value of whatever has scrap value. I wish we could offer you a better price for demolition and disposal, but there is simply no more room. I hope this explanation answers your question. Thank you for your support.

Sincerely,

(Name)
(Title)

Company Name, Address, Tel., Fax., E-mail, Web-Site Address

LETTER TYPE:	NOTICE
ADDRESSED TO:	**ARCHITECT**
RE:	**CONTRACTUAL ADDENDUM**

SCENARIO:

This letter represents a contractual addendum resulting from discrepancies between the site plan and grading plan. The contractor initially writes the following letter as an attempt to resolve the differences.

COMPANY LETTERHEAD

(Recipient's Name)   (Date)
(Recipient's Title)
(Recipient's Contact Info.)

RE: (Project's Name and Tracking Number)

Dear (Recipient's Name),

After our pre-construction meeting on (date), the civil engineer noted some discrepancies between your site plan and his grading plan. The primary issue is a conflict of property lines. We recommend that you instruct the owner to identify the correct property lines and review the site plan for inconsistencies. By standard practice, the owner shall locate and point out property lines to the contractor. The contractor has the option to require the owner to provide a licensed land surveyor's map. We cannot commence construction until this is done.

We will assist you by preparing a supplemental contractual addendum to resolve the property line discrepancies. Per the new agreement, the time for completion of work will be adjusted to allow an additional 30 days for execution. At that point the owner will have the jobsite ready for commencement of construction, and shall then give the contractor written notice to commence work. The contractor shall commence work within 10 days after such notice and shall complete the grading within the original stipulated schedule, subject to permissible delays as described in the original contract.

Please let me know if you have any additional concerns that we can resolve at this time. Thank you for your cooperation and assistance in this matter.

Respectfully,

(Name)
(Title)

Company Name, Address, Tel., Fax., E-mail, Web-Site Address

LETTER TYPE:	**REQUEST FOR INFORMATION**
ADDRESSED TO:	**ARCHITECT**
RE:	**SCOPE OF WORK**

SCENARIO:

Contractor writes a letter requesting a release of information from the architect.

REFERENCE NOTE:
Use- BNi-W Form 225 as a Notice of Request for Technical Instructions
_OR_
Use- BNi-W Form 302 as a Notice of Request for Bid if soliciting sub-bids
_OR_
Use- BNi-W Form 273 for a general proposal describing the scope of work.
_OR_
Use- BNi-W Form 113 as a Bid Confirmation describing the scope of work.

COMPANY LETTERHEAD

(Recipient's Name)   (Date)
(Recipient's Title)
(Recipient's Contact Info.)

RE: (Project's Name and Tracking Number)

Dear (Recipient's Name),

Thank you for inviting us to your pre-construction meeting. We would appreciate it very much if you could provide us with some information on the scope of work being discussed prior to the meeting.

Your assistance with this information will help us provide you with a more accurate proposal for the job. We would also like to offer you a construction phasing plan that best suits this project.

I look forward to receiving your information.

Sincerely,

(Name)
(Title)

Company Name, Address, Tel., Fax., E-mail, Web-Site Address

LETTER TYPE:	APPROVAL REQUEST
ADDRESSED TO:	ARCHITECT
RE:	CHANGE ORDER

SCENARIO:

Site conditions have triggered a construction change affecting the footing's depth on a project. After careful consideration, the geotechnical and structural engineers designed deeper footings. The architect has issued a bulletin ordering the contractor to perform the changes. The contractor prepares a change order with the following transmittal.

REFERENCE NOTE:
Use-BNi-W Form 211 as a Change Order Form for use in the performance of the contract.

OR

Use-BNi-W Form 232 as a Change Order Form if the client requires a quotation for a change in scope of work.

OR

Use-BNi-W Form 226 if you need to give Official Notice of Possible Cost Increase Confirmation.

COMPANY LETTERHEAD

(Recipient's Name)   (Date)
(Recipient's Title)
(Recipient's Contact Info.)

RE: (Project's Name and Tracking Number)

Dear (Recipient's Name),

    In reference to bulletin (number), received on (date), we are requesting a cost increase to reflect the footing depth changes. Please review the attached Change Order (number). I look forward to your approval on this before starting the formwork.

    Please feel free to call me if you have any questions or concerns.

Sincerely,

(Name)
(Title)

Company Name, Address, Tel., Fax., E-mail, Web-Site Address

LETTER TYPE: **APPROVAL REQUEST**
ADDRESSED TO: **ARCHITECT**
RE: **CHANGE ORDER**

SCENARIO:

A contractor has issued seven change orders and wants to urge the architect to review the drawings in order to avoid additional complications at the owner's expense.

REFERENCE NOTE:
Use-BNi-W Form 291 as the Notice of Disclaimer and Protest for work that can't be done.
OR

Use-BNi-W Form 320 is you need to provide the Architect with a Notice of Directive or Communication.
OR

Use-BNi-W Form 111 for an Extra Work Confirmation.
OR

Use-BNi-W Form 224 for a Notice of Acceleration if you need to speed-up work or directives.

COMPANY LETTERHEAD

(Recipient's Name)   (Date)
(Recipient's Title)
(Recipient's Contact Info.)

RE: (Project's Name and Tracking Number)

Dear (Recipient's Name),

    As you know, multiple changes always disrupt orderly work performance, causing loss of productivity and increased cost.

    Inasmuch as there have been (number) changes to date, we urge you, as the architect, to review the documents for additional errors, omissions, and previously unforeseen needs as soon as possible. We ask this so that the required changes may be resolved and executed in time for the work to be completed with minimum interference and associated costs.

    If a fair and reasonable price and time cannot be negotiated prior to performance in some situations, we will request your authorization to proceed on a time and material basis. This will be in accordance with the provisions of (insert the appropriate description of the change clause included in the specific contract).

Thank you for your understanding.

Very truly yours,

(Name)
(Title)

Company Name, Address, Tel., Fax., E-mail, Web-Site Address

LETTER TYPE:	DETERMINATION REQUEST
ADDRESSED TO:	ARCHITECT
RE:	ALTERNATIVE DISPUTE RESOLUTION

SCENARIO:

A contractor has received a demand from the client to demolish and reconstruct a portion of a project.

The demands are based on aesthetics, and not performance. The contractor opts to ignore this request since it exceeds the retainer for this project. The contractor is aware of his legal rights, and is advised by his attorney to obtain a final determination from the architect on an alternative dispute resolution in his favor.

COMPANY LETTERHEAD

(Recipient's Name)   (Date)
(Recipient's Title)
(Recipient's Contact Info.)

RE: (Project's Name and Tracking Number)

Dear (Recipient's Name),

We had the opportunity to discuss the situation that we are facing with the owner's demands for us to demolish and rebuild the co-location cages for the above referenced project. As clearly expressed by the owner, this decision is purely based on aesthetics. As we explained to you, demolishing and rebuilding the project would exceed the retainer by as much as 300%. You inspected the job and found it to be consistent with your specifications and working drawings. The inspector approved it and we resolved the punch list items per your request. We feel that the client's demands are unreasonable. At this point, we are requesting that you take action, as you hold ultimate control. We know that your determination in a dispute regarding matters of aesthetics is considered final. It is our understanding that the following are possible repercussions:

1. The owner withholding payments.

2. The owner ejecting us from the jobsite.

3. We demolish the co-location cages and equipment and take a loss.

4. We ignore the owner's request and continue with the work as scheduled.

After consulting with our attorney, we are opting to ignore the client's demand and continue with the work. If the client attempts a stop payment and terminates us, we will be forced to sue for the amount withheld. At that point, we would declare the client in default and file suit to recover costs, plus the profit we would have earned if allowed to continue.

I am sure you will agree that the best resolution for everyone would be for you to make final determination and avoid costly mediation or litigation. Please let us hear from you regarding this situation as soon as possible.

Very truly yours,

(Name)
(Title)

Company Name, Address, Tel., Fax., E-mail, Web-Site Address

LETTER TYPE: **APPROVAL REQUEST**
ADDRESSED TO: **ARCHITECT**
RE: **EQUIPMENT COORDINATION**

SCENARIO:

A contractor requests that the architect review the equipment list to make sure it will comply with any required specifications. The architect's approval is critical to the decision for ordering the equipment and establishing its schedule.

REFERENCE NOTE:
Use-BNi-W Form 225 as a Notice of Request for Technical Instructions.
OR
Use-BNi-W Form 320 if you need a Notice of Directive or Communication from the architect.

COMPANY LETTERHEAD

(Recipient's Name)   (Date)
(Recipient's Title)
(Recipient's Contact Info.)

RE: (Project's Name and Tracking Number)

Dear (Recipient's Name),

Attached is the list of mechanical and electrical equipment which has been approved to date and which will be incorporated into the project, along with their respective connections and support requirements.

Please confirm that the present design of the facility will accommodate each piece of equipment as listed. If you should discover a discrepancy, please advise this office immediately of any necessary corrective action.

Construction is progressing based on the information included in the contract documents and in accordance with the current construction schedule. To minimize the impact of any possible change at this late date, your immediate review and response is required.

Thank you for your prompt response.

Respectfully,

(Name)
(Title)

Company Name, Address, Tel., Fax., E-mail, Web-Site Address

LETTER TYPE: INSPECTION REQUEST
ADDRESSED TO: CONCRETE INSPECTOR
RE: SCHEDULE INSPECTION

SCENARIO:

A contractor has completed the footing preparation at a new residential project. He needs to obtain the footing approval before starting work on the slab. He writes an e-mail to the local city inspector who has already reviewed parts of the project.

COMPANY LETTERHEAD

(Recipient's Name)   (Date)
(Recipient's Title)
(Recipient's Contact Info.)

RE: (Project's Name and Tracking Number)

Dear (Recipient's Name),

   This is a follow-up message regarding the concrete footing inspection at (address). We expect to commence the slab work by (date). Please let me know if you will be available to perform an inspection by (date).

   Please don't hesitate to contact me if you observe any unforeseen problems with the scheduling of this inspection appointment.

   Thank you very much for your time and attention to this matter.

Cordially,

(Name)
(Title)

Company Name, Address, Tel., Fax., E-mail, Web-Site Address

LETTER TYPE: **ARGUMENTATIVE**
ADDRESSED TO: **FRAMING INSPECTOR**
RE: **ARGUMENT AGAINST A FAILED INSPECTION**

SCENARIO:

A framing inspector has denied approval on the framing of a new building addition project. The contractor understands that all pre-existing conditions should not be a denial for approvals of new work, with the exception of ADA regulations. The contractor decides to challenge the inspection report.

COMPANY LETTERHEAD

(Recipient's Name)   (Date)
(Recipient's Title)
(Recipient's Contact Info.)

RE: (Project's Name and Tracking Number)

Dear (Recipient's Name),

    It came as a surprise that your framing inspection was greatly influenced by existing conditions. We understand such conditions are considered "grandfathered," and should not interfere with the approval of our new project. Your inspection did not find any code deficiencies or violations on our new work.

    At this time, we are requesting that you provide us with the code citation you used to deny the "grandfather clauses" that protect our new work from existing conditions. I would greatly appreciate it if you could answer this request at your earliest possible convenience. Thank you in advance for your prompt response.

Cordially,

(Name)
(Title)

Company Name, Address, Tel., Fax., E-mail, Web-Site Address

LETTER TYPE:	DISAPPROVAL
ADDRESSED TO:	**BUILDING INSPECTOR**
RE:	**AIR-LEAKAGE INSPECTION**

SCENARIO:

An inspector has provided the contractor with a *Failed Inspection Card* on the air pressure testing done for a project.

The contractor for this project contacts the manufacturer to make sure the product has been installed correctly, and functions according to its warranty and performance specifications.

After a company service representative inspects each door and performs random air-pressure testing, the contractor determines that the product's performance is satisfactory. He decides to challenge the inspection by writing the following letter.

COMPANY LETTERHEAD

(Recipient's Name)   (Date)
(Recipient's Title)
(Recipient's Contact Info.)

RE: (Project's Name and Tracking Number)

Dear (Recipient's Name),

The air leakage inspection for the above referenced project failed according to your readings and determination. At this point, we are contesting the findings of your inspection and would like to schedule a re-inspection at your earliest convenience.

We asked the manufacturer to perform the same air leakage test on the subject spaces. All doors passed the test by achieving a negative pressure, with respect to the adjacent spaces of an average 5 Pa. We have documented these results and would like to review them with you.

Thank you for your time and assistance.

Cordially,

(Name)
(Title)

Company Name, Address, Tel., Fax., E-mail, Web-Site Address

LETTER TYPE:	APPRECIATIVE
ADDRESSED TO:	BUILDING INSPECTOR
RE:	THANK YOU LETTER

SCENARIO:

A contractor writes a follow-up letter to thank the inspector of his recently completed project.

COMPANY LETTERHEAD

(Recipient's Name)   (Date)
(Recipient's Title)
(Recipient's Contact Info.)

RE: (Project's Name and Tracking Number)

Dear (Recipient's Name),

   I am writing to thank you for the professional inspection done on my project. Your ability and field experience assisted us tremendously by making us aware of the new code regulations. Your kind and professional disposition made this experience a memorable one.

   We look forward to working with you on the (project name) scheduled for (date).

   Thank you once again.

Best Regards,

(Name)
(Title)

Company Name, Address, Tel., Fax., E-mail, Web-Site Address

LETTER TYPE:	**REQUEST**
ADDRESSED TO:	**BUILDING INSPECTOR**
RE:	**INSPECTION REQUEST**

SCENARIO:

A contractor has reviewed the inspection log regarding the storm drains for a major development project. The contractor has completed the tasks involved with the log. This letter reviews the action items noted by the inspector and determines completion of them. This letter also requests a response from the inspector in order to accelerate project close-out.

COMPANY LETTERHEAD

(Recipient's Name)   (Date)
(Recipient's Title)
(Recipient's Contact Info.)

RE: (Project's Name and Tracking Number)

Dear (Recipient's Name),

Per your request, we have cleaned all catch basins and flushed all storm drain lines prior to job acceptance.

You have inspected the interior of all drainage structures for rock pockets and defects in workmanship. We provided you with the required permit for tying in the site drainage storm drain lines into the jurisdiction's main lines. You inspected all storm drain lines for joint integrity, that lines are true for slope and horizontal alignment, and that bedding is proper prior to allowing backfill.

Please let us know if there are any outstanding punch list items.

Thank you for your assistance.

Sincerely,

(Name)
(Title)

Company Name, Address, Tel., Fax., E-mail, Web-Site Address

LETTER TYPE: **REPRIMAND**
ADDRESSED TO: **BUILDING INSPECTOR**
RE: **NEGLIGENCE**

SCENARIO:

The superintendent of a project was injured by poorly installed electrical cables. This accident could have been avoided if the inspector reported the conditions to the construction manager or superintendent. The inspector recently completed an electrical inspection and noted the deficiency on his report. However, he did not warn the superintendent of the potential hazard before releasing the report.

COMPANY LETTERHEAD

(Recipient's Name)   (Date)
(Recipient's Title)
(Recipient's Contact Info.)

RE: (Project's Name and Tracking Number)

Dear (Recipient's Name),

We are seriously concerned about your failure to inform the superintendent of the potential hazards you encountered at the jobsite while performing your routine inspection. The result was an accident that could have been easily avoided.

We understand that the construction inspector is not directly responsible for public safety. However, unsafe conditions cannot be ignored, and codes and regulations are typically developed in order to ensure public health, safety and welfare.

If the construction inspector observes anything which might be unsafe, it should be immediately reported to the contractor's superintendent. If a member of the general public might be in danger, it would be appropriate for the construction inspector to stay and do whatever is possible to prevent a potential accident.

Please inform us immediately of any such conditions in the future.

Sincerely,

(Name)
(Title)

Company Name, Address, Tel., Fax., E-mail, Web-Site Address

LETTER TYPE:	ENVIRONMENTAL
ADDRESSED TO:	**COUNTY HAZMAT OFFICIALS**
RE:	**WASTE DIVERSION**

SCENARIO:

A demolition contractor has received a voluntary notice from the county hazmat department to participate in the latest "Waste Management Incentive Program." The contractor is happy to participate and responds to this invitation:

COMPANY LETTERHEAD

(Recipient's Name)   (Date)
(Recipient's Title)
(Recipient's Contact Info.)

RE: (Project's Name and Tracking Number)

Dear (Recipient's Name),

    We received your "Waste Management Incentive Program" invitation and are more than happy to participate. In our construction waste management plan, we have specified a 75% diversion from disposal. This guideline was enforced by the architect's LEED AP, according to MR Credit 2.1. Through this effort, we are redirecting reusable materials to appropriate sites. In addition, our report identifies which demolition debris items are traced and delivered directly to their resource manufacturers.

    Whenever possible, we have included donation items for charity purposes. Our intent is to respond to your latest environmental policies. For full compliance, we will schedule an inspection shortly.

Respectfully,

(Name)
(Title)

Company Name, Address, Tel., Fax., E-mail, Web-Site Address

LETTER TYPE: **ENVIRONMENTAL**
ADDRESSED TO: **STATE ENVIRONMENTAL**
RE: **COMMISSION**
                      OZONE DEPLETION REDUCTION

SCENARIO:

A general contractor is applying for an incentive to be applied towards the purchase and use of energy-efficient products.

COMPANY LETTERHEAD

(Recipient's Name)   (Date)
(Recipient's Title)
(Recipient's Contact Info.)

RE: (Project's Name and Tracking Number)

Dear (Recipient's Name),

    Please find the enclosed incentive plan we are using for the retrofit of the above referenced office building.

    We understand that compliance with these new regulations will provide federal and state incentives of up to 30% of material costs (for "E" products).

    In the report, we listed which products and strategies apply. (list description here i.e., As an example, where mechanical cooling is used, a refrigerant without HFCS or Halon was used to eliminate ozone depletion factors. New triple-pane windows have replaced the old single-pane existing windows. We have also installed foam insulation on the walls and ceiling cavities achieving a value of R-30.)

    For a more comprehensive list of "E" products, please refer to the report. If possible, please let us know when we can expect disbursement of the funds.

    Please let me know if you have any questions regarding our submittal package.

    Thank you for your time.

Sincerely,

(Name)
(Title)

Company Name, Address, Tel., Fax., E-mail, Web-Site Address

LETTER TYPE: PROJECT PERMIT REQUEST
ADDRESSED TO: BUILDING DEPARTMENT OFFICIAL
RE: PROJECT PERMITS

SCENARIO:

An architect for a major remodel informed the contractor that the plans are ready to be permitted. The contractor decides to schedule a pick-up time with the building official who assisted with this project:

COMPANY LETTERHEAD

(Recipient's Name)   (Date)
(Recipient's Title)
(Recipient's Contact Info.)

RE: (Project's Name and Tracking Number)

Dear (Recipient's Name),

We understand that the architect for Project Tracking #(number) has successfully completed the project correction process you requested.

At this time, we would like to schedule a project permit meeting with you. Please let us know of your earliest availability. If possible, please e-mail me the exact permit amount and contractor requisites (workers' comp., etc.).

Thank you for your time and assistance.

Respectfully,

(Name)
(Title)

Company Name, Address, Tel., Fax., E-mail, Web-Site Address

LETTER TYPE: **PERMIT EXTENSION REQUEST**
ADDRESSED TO: **BUILDING/TRAFFIC OFFICIAL**
RE: **EXPIRED STREET CLOSURE PERMIT**

SCENARIO:

A contractor is forced to request an extension on a street closure permit due to unexpected site conditions.

COMPANY LETTERHEAD

(Recipient's Name)   (Date)
(Recipient's Title)
(Recipient's Contact Info.)

RE: (Project's Name and Tracking Number)

Dear (Recipient's Name),

I am aware that our Street Closure Permit #(number) expired (date). We are experiencing some difficulties due to existing site conditions. Specifically, the existing infrastructure is outdated and partially destroyed. We are currently working with the local gas and electric companies in order to expedite the construction process. At this time, we are forced to request for an extension to our current Street Closure Permit.

Please let me know if you find it necessary to meet in person in order for us to facilitate this extension. Don't hesitate to contact me if you have any concerns regarding this request.

Thank you for your consideration and assistance.

Very truly yours,

(Name)
(Title)

Company Name, Address, Tel., Fax., E-mail, Web-Site Address

LETTER TYPE: **CLARIFICATION**
ADDRESSED TO: **SUBCONTRACTOR**
RE: **RESPONSIBILITY DETERMINATION**

SCENARIO:

Contractor writes a letter to subcontractor regarding the owner's decision.

REFERENCE NOTE:
Use-BNi-W Form 204 for a Notice of Disclaimer and Protest on job related disagreements.

OR

Use-BNi-W Form 205 if you need a Release of Responsibility for work to be or that has been performed.

OR

Use-BNi-W Form 225 as a Notice of Request for Technical Instructions.

COMPANY LETTERHEAD

(Recipient's Name)   (Date)
(Recipient's Title)
(Recipient's Contact Info.)

RE: (Project's Name and Tracking Number)

Dear (Recipient's Name),

Attached is the (date) (owner's name) determination of (subcontractor) responsibility for the subject work.

The work is required to be complete by (date) to avoid schedule interference and associated charges. Accordingly, please respond immediately with the following information:

1. Confirm acceptance,

2. Advise this office by (date) of your proposed schedule for completing this work.

Very truly yours,

(Name)
(Title)

Company Name, Address, Tel., Fax., E-mail, Web-Site Address

LETTER TYPE: NOTICE TO TAKE ACTION
ADDRESSED TO: SUBCONTRACTOR
RE: 48-HOUR DEFICIENCY CORRECTION

SCENARIO:

Contractor writes a letter to subcontractor regarding the completion of work described under the subcontract. This letter is intended to legally pressure the subcontractor to finish or correct any deficiency within 48 hours. It is necessary that the contractor submits this letter and keeps it as a record in case litigation results.

REFERENCE NOTE:
Use-BNi-W Form 218 as the 48-Hour Notice to Correct Deficiency in Workmanship. This form itemizes all items to be corrected or completed by the subcontractor.

COMPANY LETTERHEAD

(Recipient's Name)   (Date)
(Recipient's Title)
(Recipient's Contact Info.)

RE: (Project's Name and Tracking Number)

Dear (Recipient's Name),

   This letter has been sent to you, or a responsible employee of yours, to correct items as noted on the 48-Hour Notice to Correct Deficiencies form attached to this letter. Failure on your part to do said work will be interpreted by us as refusal, and we will have the work done AT YOUR EXPENSE by any means available to us, without regard for cost of said repair or further notice of any kind.

   If any correction notices issued to you are outstanding as of the date when you expect to draw against moneys due you, this is your notice that NO PAYMENT OF FUNDS will be made to you unless and until you have done all corrective work on orders issued to you during the month preceding expected payment.

   The person issuing this correction notice is the only person who can release funds due you. Therefore, it is in your interest to return the signed correction slip to the person who issued it to you, certifying that the correction is done. For a full list of deficient items to be corrected, please see the form attached.

   Please let us hear from you at your earliest convenience so that we may resolve this issue. Thank you for your cooperation.

Respectfully,

(Name)
(Title)

Company Name, Address, Tel., Fax., E-mail, Web-Site Address

LETTER TYPE: CORRECTIVE
ADDRESSED TO: SUBCONTRACTOR
RE: JOB COORDINATION

SCENARIO:

This letter is given to a subcontractor as a *"Corrective Notice"* in order to resolve coordination conflicts between the subcontractor's work and other trades involved in the job. This letter can also be used as a courtesy or warning letter prior to actions due to negligence.

REFERENCE NOTE:
Use-BNi-W Form 204 for a Notice of Disclaimer and Protest on job related disagreements.

OR

Use-BNi-W Form 205 if you need a Release of Responsibility for work to be or that has been performed.

OR

Use-BNi-W Form 225 as a Notice of Request for Technical Instructions.

COMPANY LETTERHEAD

(Recipient's Name)   (Date)
(Recipient's Title)
(Recipient's Contact Info.)

RE: (Project's Name and Tracking Number)

Dear (Recipient's Name),

In accordance with General Conditions Article (insert number of the pass-through clause and any other direct reference) and paragraph (insert number) of your subcontract you are required to coordinate your work with the work of all other trades.

As you may already know, the spaces above the ceilings are very restricted. Close and timely coordination of all building systems is therefore critical to avoid unnecessary interference. To date, there have been several conflicts discovered in the ceiling spaces. You are therefore cautioned to review all work in the ceiling spaces in detail, and confirm to this office all potential conflicts you identify.

Your response by (date) is required to minimize impact and delay of any conflict that may interfere with this process. Please be advised that failure to respond by (date) may leave you responsible for costs resulting from lack of coordination.

Thank you for your cooperation.

Very truly yours,

(Name)
(Title)

Company Name, Address, Tel., Fax., E-mail, Web-Site Address

LETTER TYPE: CHANGE ORDER
ADDRESSED TO: SUBCONTRACTOR
RE: COORDINATION

SCENARIO:

Contractor writes a letter to subcontractor regarding coordination of work on a project's ceiling space.

REFERENCE NOTE:
Use-BNi-W Form 204 for a Notice of Disclaimer and Protest on job related disagreements.

OR

Use-BNi-W Form 205 if you need a Release of Responsibility for work to be or that has been performed.

OR

Use-BNi-W Form 225 as a Notice of Request for Technical Instructions.

COMPANY LETTERHEAD

(Recipient's Name)   (Date)
(Recipient's Title)
(Recipient's Contact Info.)

RE: (Project's Name and Tracking Number)

Dear (Recipient's Name),

General Conditions (or other appropriate reference) Article ( ) requires that you call attention to any and all differences and deviations in the materials and equipment to be provided under your subcontract no. ( ) in their respective submittals. This requirement applies whether the items in question are submitted as specified, as equal, or as substitutions.

Your approval of submissions are required on or before the dates indicated by the current construction schedule. It is our responsibility to be aware of the current schedule, and to comply with it in every respect.

Please be advised that your failure to submit shop drawings in proper form and in a timely manner, as well as your failure to bring all deviations directly to the attention of (insert name and your own company) and the architect, may result in interference, disruption, and delay. Your company will be charged for all costs which result from failure to comply with your contractual obligations. Thank you for your cooperation.

Very truly yours,

(Name)
(Title)

Company Name, Address, Tel., Fax., E-mail, Web-Site Address

LETTER TYPE:	**CONTRACTUAL NOTICE**
ADDRESSED TO:	**SUBCONTRACTOR**
RE:	**SUBSTITUTION AGREEMENT**

SCENARIO:

Contractor writes a letter to provide a substitution agreement after leaving the site due to a major disagreement with the owner. The contractor wants to make sure he is protected by law when passing the responsibility to another contractor to finish the job. A revised contract agreement is attached to the letter. The letter and the revised contract represent the official release of liability the contractor needs to establish with all parties. The letter and revised contract can be supplemented with the forms listed below.

> **_REFERENCE NOTE:_**
> _Use-BNi-W Form 313 for a Substitution Agreement (owner, general contractor, and subcontractor)._
> OR
> _Use-BNi-W Form 314 if the Substitution Agreement is between the owner and a materialman._
> OR
> _Use-BNi-W Form 315 if the Substitution Agreement is between a general contractor and a subcontractor._

COMPANY LETTERHEAD

(Recipient's Name)   (Date)
(Recipient's Title)
(Recipient's Contact Info.)

RE: (Project's Name and Tracking Number)

Dear (Recipient's Name),

I regret the outcome between the client and myself, which forced us to terminate our existing contract. Regardless of the level of discomfort, however, I will do my best to resolve this situation with professionalism and responsibility.

I release all responsibility and liability placed on me for this job by adjusting the original sub-contractual agreement. A copy of this subcontract is attached, and you, the subcontractor, and the owner will agree to substitute me, the general contractor, in the subcontract. It is further agreed that all other terms and conditions of the original subcontract, including payment, price and completion, shall remain in full force and in effect without change, and the owner shall have all rights and powers under the attached subcontract which the general contractor had. By this, I relinquish any further liability and responsibility.

Please review the subcontract in detail and make sure all provisions have been established according to both parties. Please feel free to contact me at your earliest convenience if you feel we need to discuss the subcontract provisions prior to your signature determination.

Sincerely,

(Name)
(Title)

Company Name, Address, Tel., Fax., E-mail, Web-Site Address

LETTER TYPE: CHANGE ORDER
ADDRESSED TO: SUBCONTRACTOR
RE: QUOTATION FOR ADDITIONAL WORK

SCENARIO:

Contractor's letter to subcontractor regarding a proposal request.

REFERENCE NOTE:
Use-BNi-W Form 112 for a Conversation Confirmer that can be faxed for
a written approval of a quotation.

OR

Use-BNi-W Form 318 when releasing a Notice of Request for Change Order with
Quotation.

OR

Use-BNi-W Form 312 for a Notice of Request for Change in Specifications and
Substitutions.

OR

Use-BNi-W Form 313 if the Owner and Subcontractor need a Substitution
Agreement

OR

Use-BNi-W Form 319 for a Notice of Request for Price.

COMPANY LETTERHEAD

(Recipient's Name)   (Date)
(Recipient's Title)
(Recipient's Contact Info.)

RE: (Project's Name and Tracking Number)

Dear (Recipient's Name),

Attached is the (identify and date all enclosures necessary to price the change). If the subject work does not affect your trade, please submit your confirmation of no change in contract price or time. If the changed work affects your company, please submit the following:

1. Cost to complete the changed work,

2. Any applicable credit for contract work,

3. All substantiating labor and material records,

4. Material and equipment delivery times after change order approval,

5. Time required to perform the work (separate major items),

6. Work of any other trade affected,

7. All conditions required to perform the work,

8. Any significant weather, site, or other constraints beyond your control,

9. All other applicable information.

Your attention is called to (insert the appropriate subcontract article reference) for proper format and required level of detail. If you have any questions, please contact me immediately.

Your complete response, in proper form, is required by (date) to avoid unnecessary delay associated with this change.

Very truly yours,

(Name)
(Title)

Company Name, Address, Tel., Fax., E-mail, Web-Site Address

LETTER TYPE: **CHANGE ORDER**
ADDRESSED TO: **SUBCONTRACTOR**
RE: **QUOTATION REQUEST, NON RESPONSE**

SCENARIO:

Contractor writes a letter to subcontractor requesting a change order quotation. The subcontractor has not responded to the previous correspondence.

REFERENCE NOTE:
Use-BNi-W Form 112 for a Conversation Confirmer that can be faxed for a written approval of a quotation.
OR
Use-BNi-W Form 318 when releasing a Notice of Request for Change Order with Quotation.
OR
Use-BNi-W Form 312 for a Notice of Request for Change in Specifications and Substitutions.
OR
Use-BNi-W Form 313 if the Owner and Subcontractor need a Substitution Agreement
OR
Use-BNi-W Form 319 for a Notice of Request for Price.

COMPANY LETTERHEAD

(Recipient's Name)   (Date)
(Recipient's Title)
(Recipient's Contact Info.)

RE: (Project's Name and Tracking Number)

Dear (Recipient's Name),

The following requests for change quotations remain outstanding:

Change No. Description   Date Requested   Date Due

Your lack of response by the date required is now interfering with the completion of our proposal and with the work. It is essential that your proposal(s) is (are) now to be delivered to this office by (allow two days). In the event we do not receive proper response, in the form required by your subcontract, your company will be charged for all costs associated with any resulting delay.

Please take notice that we reserve all rights to claim all damages resulting from your untimely response.

Very truly yours,

(Name)
(Title)

Company Name, Address, Tel., Fax., E-mail, Web-Site Address

LETTER TYPE: CHANGE ORDER
ADDRESSED TO: SUBCONTRACTOR
RE: PRICE BY DEFAULT

SCENARIO:

Contractor writes a letter to subcontractor requesting quotations on outstanding change orders.

REFERENCE NOTE:
Use-BNi-W Form 112 for a Conversation Confirmer that can be faxed for a written approval of a quotation.

COMPANY LETTERHEAD

(Recipient's Name)   (Date)
(Recipient's Title)
(Recipient's Contact Info.)

RE: (Project's Name and Tracking Number)

Dear (Recipient's Name),

   Your continued failure to respond to repeated requests for change order quotations is now delaying project close-out and generating unnecessary and excessive overhead expense. Accordingly, if your price is not received by (date), a unilateral change order will be processed in the amount of ($ amount). At that time, the file will be closed, and there will be no opportunity for further review.

   Be advised that any damages resulting from your lack of attention to this matter will be back-charged to your account. Additionally, we reserve the right to claim all damages resulting from your lack of response.

   Please take notice.

Respectfully,

(Name)
(Title)

Company Name, Address, Tel., Fax., E-mail, Web-Site Address

LETTER TYPE:	CHANGE ORDER
ADDRESSED TO:	**SUBCONTRACTOR**
RE:	**TELEPHONE QUOTE**

SCENARIO:

Contractor writes a letter to subcontractor regarding a telephone quote confirmation.

REFERENCE NOTE:
Use-BNi-W Form 112 for a Conversation Confirmer that can be faxed for a written approval of a quotation.

COMPANY LETTERHEAD

(Recipient's Name)   (Date)
(Recipient's Title)
(Recipient's Contact Info.)

RE: (Project's Name and Tracking Number)

Dear (Recipient's Name),

Per our agreement of (date of the telephone quote), attached is the subject Change Order Telephone Quotation Form. Please confirm that all information set forth on this form is accurate by signing below and returning this letter to my attention.

Your response is requested by (date).

Respectfully,

(Name)
(Title)

Company Name, Address, Tel., Fax., E-mail, Web-Site Address

LETTER TYPE: CHANGE ORDER
ADDRESSED TO: SUBCONTRACTOR
RE: IMPROPER QUOTATION SUBMITTAL

SCENARIO:

Contractor writes a letter to subcontractor regarding an improper proposal submission.

REFERENCE NOTE:
Use-BNi-W Form 204 for a Notice of Disclaimer and Protest on job related disagreements.

OR

Use-BNi-W Form 205 if you need a Release of Responsibility for work to be or that has been performed.

OR

Use-BNi-W Form 110 for a Contractor's Declaration to Procure Payment.

COMPANY LETTERHEAD

(Recipient's Name)   (Date)
(Recipient's Title)
(Recipient's Contact Info.)

RE: (Project's Name and Tracking Number)

Dear (Recipient's Name),

Per our conversation earlier today, your proposal for the subject change order is being returned for correction. Article (insert appropriate contract or subcontract provision) requires a detailed breakdown, properly itemized to allow evaluation.

Please correct your proposal and resubmit in the proper format. The corrected submission is required by (date of two working days after conversation date). Your lack of response after that date will result in interferences that will be your responsibility.

Please be advised that any excessive time required to evaluate and process improper proposal submissions will be back-charged to your company, and that you will be held responsible for delays and additional interferences resulting from improper action.

Please correct and resubmit your proposal immediately.

Very truly yours,

(Name)
(Title)

Company Name, Address, Tel., Fax., E-mail, Web-Site Address

LETTER TYPE: **CHANGE ORDER**
ADDRESSED TO: **SUBCONTRACTOR**
RE: **TIME AND MATERIALS**

SCENARIO:

This letter is addressed to the subcontractors regarding time and materials tickets. The general contractor is requesting the subcontractors to follow the T&M ticket procedures in order to achieve timely payments. This process needs to be constantly monitored by the general contractor's staff in order to ensure that the work conforms to the contract prior to the release of any payment due to the subcontractors.

.

REFERENCE NOTE:

Use-BNi-W Form 318 when releasing a Notice of Request for Change Order with Quotation.

OR

Use-BNi-W Form 312 for a Notice of Request for Change in Specifications and Substitutions.

OR

Use-BNi-W Form 313 if the Owner and Subcontractor need a Substitution Agreement

OR

Use-BNi-W Form 319 for a Notice of Request for Price.

COMPANY LETTERHEAD

(Recipient's Name)   (Date)
(Recipient's Title)
(Recipient's Contact Info.)

RE: (Project's Name and Tracking Number)

Dear (Recipient's Name),

On (date), your company was directed to proceed with the subject work on a time and material basis, in accordance with the provisions (insert appropriate contract or subcontract reference). The conditions of this arrangement are as follows:

1. T&M tickets are to be signed daily. Tickets that are not signed on the day the work was actually performed will not be recognized as additional expense items.

2. Precise labor classifications are to be noted on each ticket.

3. Material invoices are to be attached.

Signatures by field personnel only confirm that certain work was performed with certain forces on that day. Field personnel do not acknowledge or agree that the work is in addition to the contract or that the rates charged are acceptable. Both of these matters are subject to further review in accordance with the terms (insert appropriate contract or subcontract references).

Thank you for your cooperation.

Respectfully,

(Name)
(Title)

Company Name, Address, Tel., Fax., E-mail, Web-Site Address

LETTER TYPE: INVITATION
ADDRESSED TO: SUBCONTRACTOR
RE: MANDATORY MEETING ATTENDANCE

SCENARIO:

Contractor writes a letter to subcontractor requesting attendance at mandatory job meetings.

REFERENCE NOTE:
Use-BNi-W Form 203 for Speed Memos when communicating with other firms and organizations.

OR

Use-BNi-W Form 203A for internal Speed Memos if you need a record reminder of the meeting.

OR

Use-BNi-W Form 212 for final Meeting Minutes Reporting.

COMPANY LETTERHEAD

(Recipient's Name)   (Date)
(Recipient's Title)
(Recipient's Contact Info.)

RE: (Project's Name and Tracking Number)

Dear (Recipient's Name),

Your subcontract requires your participation in regular job meetings. These meetings have been scheduled to begin (date) and will be held on alternating (day of the week). During critical or problem periods, the meetings may be held weekly, as determined by our company. It is your responsibility to be aware of the current job meeting schedule.

(Names) are required to attend all meetings. All other subcontractors performing or about to perform work on the site are required to attend all meetings throughout the period of their work.

Please note that this is not a request. Your attendance at these meetings is mandatory. Your failure to attend job meetings will result in extra efforts by others to coordinate their work with your work. Because these extra efforts disrupt the progress of the work and inconvenience other tradespersons on this project, absences cannot be tolerated.

Please be advised that you will be held responsible for all information contained in the meeting minutes, including timetables, commitments, and determinations of responsibility as set forth in the minutes.

Thank you for your cooperation.

Very truly yours,

(Name)
(Title)

Company Name, Address, Tel., Fax., E-mail, Web-Site Address

LETTER TYPE: REPRIMAND
ADDRESSED TO: SUBCONTRACTOR
RE: LACK OF JOB MEETING ATTENDANCE

SCENARIO:

Contractor writes a letter to subcontractor for missing the jobsite meeting.

REFERENCE NOTE:
Use-BNi-W Form 203 for Speed Memos when communicating with other firms and organizations.

OR

Use-BNi-W Form 203A for internal Speed Memos if you need a record reminder of the meeting.

OR

Use-BNi-W Form 212 for final Meeting Minutes Reporting.

COMPANY LETTERHEAD

(Recipient's Name)   (Date)
(Recipient's Title)
(Recipient's Contact Info.)

RE: (Project's Name and Tracking Number)

Dear (Recipient's Name),

    As we discussed earlier today, your failure to attend today's job meeting as required is interfering with job coordination and completion. As you already know, it is your responsibility to be aware of all project requirements and to accommodate them completely and in a timely manner. Please be advised that you will be held responsible for all interferences, delays, and added costs resulting from your lack of attention to these requirements.

    The next job meeting will be held on (insert day and date) promptly at (insert time).

Very truly yours,

(Name)
(Title)

Company Name, Address, Tel., Fax., E-mail, Web-Site Address

LETTER TYPE: LEGAL AGREEMENT DISCLOSURE
ADDRESSED TO: SUBCONTRACTOR
RE: HOLD HARMLESS AGREEMENT

SCENARIO:

A general contractor failed to receive hold harmless coverage from the concrete subcontractor of a building. The GC arranged for a meeting to resolve this issue, and decided to follow up by faxing or mailing the official Hold Harmless Form with the following cover sheet message.

REFERENCE NOTE:
Use-BNi-W Form 249 for an Indemnity and Hold Harmless Agreement.

COMPANY LETTERHEAD

(Recipient's Name)   (Date)
(Recipient's Title)
(Recipient's Contact Info.)

RE: (Project's Name and Tracking Number)

Dear (Recipient's Name),

Please find the enclosed Hold-Harmless Agreement we discussed earlier today and review and sign the appropriate spaces. As previously discussed under the subcontract/provisions for the project, you need to provide hold harmless protection to the general contractor for work done under your contract. You, as the undersigned, shall protect, hold free and harmless, defend and indemnify indemnitee, its directors, officers, agent and employees from and against any and all claims, debts, demands, damages (including direct, liquidated, consequential, incidental, or other damages), judgments, awards, losses, liabilities, interest, expert witness fees, attorney fees, and costs and expenses of whatsoever kind or nature.

I am sure you have been asked to perform this prerequisite under different contracts you dealt with in the past. If you should have any difficulty or concern regarding the provisions noted on the Hold Harmless Form, please feel free to call me, or if you wish you may consult with your attorney. As mentioned, this is standard procedure on any job.

I appreciate your time and cooperation on this matter. Please note that you have five working days to comply by signing and returning the form.

Respectfully,

(Name)
(Title)

Company Name, Address, Tel., Fax., E-mail, Web-Site Address

LETTER TYPE: **INVITATION**
ADDRESSED TO: **SUBCONTRACTOR**
RE: **NOTICE INVITING BIDS**

SCENARIO:

A contractor sets out the requirements of subcontractors for a commercial project.

REFERENCE NOTE:
Use-BNi-W Form 301 as a general Notice for Inviting Bids
OR
Use-BNi-W Form 302 for a Notice of Request for Bid which describes in more detail the scope of work.

COMPANY LETTERHEAD

(Recipient's Name)   (Date)
(Recipient's Title)
(Recipient's Contact Info.)

RE: (Project's Name and Tracking Number)

Dear (Recipient's Name),

    A jobsite visit will be available for prospective bidders on (date) at (time) for the purpose of acquainting all prospective bidders with the bid documents and the jobsite. It is mandatory for all bidders to attend this conference; failure to attend the pre-bid conference will disqualify a prospective bidder from the project. Bids must be submitted for the entire (specify trade) scope of work described. Deviations from plans and specifications will not be considered and will cause rejection of bids. Each bid shall be made using the bid form included in the contract documents. Each bid must conform, and be responsive to this invitation, the plans and specifications, and all other documents.

    Performance and payment bonds will be required of the successful bidder within 10 business days after the notification of the award of the contract.

    We hope this provides all necessary acknowledgements. Please don't hesitate to contact us if you should have any additional questions.

Sincerely,

(Name)
(Title)

Company Name, Address, Tel., Fax., E-mail, Web-Site Address

LETTER TYPE: INFORMATIVE
ADDRESSED TO: SUBCONTRACTOR
RE: SUBCONTRACT OPERATIONS

SCENARIO:

The subcontractor of the previous letter won the bid for the commercial project. After reviewing and signing the contract, the contractor's secretary in charge of record keeping, sends the following letter to the subcontractor.

REFERENCE NOTE:
Use-BNi-W Form 305 if you need to authorize a Subcontractor or Vendor Bid.
OR
Use-BNi-W Form 102 for a Construction Subcontract
OR
Use-BNi-W Form 307 if you need a Bond Request on your Subcontract.

COMPANY LETTERHEAD

(Recipient's Name)   (Date)
(Recipient's Title)
(Recipient's Contact Info.)

RE: (Project's Name and Tracking Number)

Dear (Recipient's Name),

Enclosed are three copies of our written subcontract agreement signed by (name of authorized person). As soon as possible and before starting any work, kindly furnish us with three copies of each certificate of insurance covering worker's compensation, public liability, and property damage. We suggest that you carefully note the required limits.

In accordance with the provisions of the signed contract, you are required to provide a weekly copy of payrolls, a signed statement of compliance certifying that payrolls are correct and complete, and a fringe benefit statement indicating that the fringe benefits have been paid to the employee or the appropriate organization.

In addition, please review the specifications and forward at the earliest possible date three copies of any submittals required, such as shop drawings, catalog cuts, materials lists, and samples.

Please let me know if you have any questions or concerns. We look forward to working closely with you and your staff. Thank you for your cooperation.

Sincerely,

(Name)
(Title)

Company Name, Address, Tel., Fax., E-mail, Web-Site Address

LETTER TYPE: **BID ITEM CLARIFICATION**
ADDRESSED TO: **CONTRACTOR**
RE: **BID ITEM QUESTION**

SCENARIO:

A subcontractor has successfully won the bid for a job. He contacted the general contractor in charge to request further clarifications on the bond required for this project.

REFERENCE NOTE:
Use- BNi-W Form 225 as a Notice of Request for Technical Instructions
OR
Use- BNi-W Form 302 as a Notice of Request for Bid if soliciting sub-bids
OR
Use- BNi-W Form 273 for a general Proposal describing the scope of work.
OR
Use- BNi-W Form 113 as a Bid Confirmation describing the scope of work.

COMPANY LETTERHEAD

(Recipient's Name)   (Date)
(Recipient's Title)
(Recipient's Contact Info.)

RE: (Project's Name and Tracking Number)

Dear (Recipient's Name),

    I am writing you to inform you that we have been selected as the subcontractor of choice for the mechanical project (number). We spoke this morning regarding the bond required for this job as well as the completion date.

    Could you please send this information to me in writing for my records?

    I appreciate your assistance and prompt response.

Sincerely,

(Name)
(Title)

Company Name, Address, Tel., Fax., E-mail, Web-Site Address

LETTER TYPE:	**DOCUMENT REQUEST**
ADDRESSED TO:	**SUBCONTRACTOR**
RE:	**INDEMNIFICATION REQUEST**

SCENARIO:

A general contractor has awarded a subcontractor a job. After careful contract review, the contractor realized he may be liable since an indemnification agreement has not been included with the contract. The contractor requests for this agreement to be part of the contract:

REFERENCE NOTE:
Use-BNi-W Form 102 for a Construction Subcontract (Between General and Sub).
OR
Use-BNi-W Form 326 for Labor Release on subcontractor's payment obligations
OR
Use-BNi-W Form 329 for a Notice for Compliance if a subcontractor fails to comply with the requirements of the contractual documents.
OR
Use-BNI-W Form 222 for a Notice of Release of Responsibility for work to be or that has to be performed. (adverse conditions).

COMPANY LETTERHEAD

(Recipient's Name)   (Date)
(Recipient's Title)
(Recipient's Contact Info.)

RE: (Project's Name and Tracking Number)

Dear (Recipient's Name),

I am writing to request that you provide an indemnification agreement with your contract. Please identify myself and the owner on the hold harmless clause from all loss and damage, and against all lawsuits, arbitrations, mechanic's liens, legal actions of any kind, attorney's fees, and any costs and expenses, which are directly or indirectly caused by, or contributed to, by the subcontractor (identify yourself) or his agents, employees, or sub-subcontractors, in connection with or incidental to, the work under this agreement.

This agreement will be reviewed and signed by the owner and myself prior to your scheduled commencement day. Please contact me if you have any questions or concerns.

Thank you for your cooperation.

Sincerely,

(Name)
(Title)

Company Name, Address, Tel., Fax., E-mail, Web-Site Address

LETTER TYPE: **BID ITEM CLARIFICATION**
ADDRESSED TO: **SUBCONTRACTOR**
RE: **BID ITEM QUESTION**

SCENARIO:

The general contractor responds to the successful bidder of the previous letter.

REFERENCE NOTE:
Use- BNi-W Form 225 as a Notice of Request for Technical Instructions
OR
Use- BNi-W Form 302 as a Notice of Request for Bid if soliciting sub-bids
OR
Use- BNi-W Form 273 for a general Proposal describing the scope of work.
OR
Use- BNi-W Form 113 as a Bid Confirmation describing the scope of work.

COMPANY LETTERHEAD

(Recipient's Name)   (Date)
(Recipient's Title)
(Recipient's Contact Info.)

RE: (Project's Name and Tracking Number)

Dear (Recipient's Name),

Per our conversation earlier today, all successful bidders will be required to furnish a performance bond and a labor and material bond in the statutory form of public bonds required by Sections (number) and (number) of the State Finance Law. The amount of the contract is estimated to be between ($ amount) and ($ amount). Performance bonds may be waived on a bid under ($ amount). The completion date for this project is (number) days after the agreement is approved.

I hope this information addresses your concerns. Please don't hesitate to contact me if you should have any additional questions.

Respectfully,

Company Name, Address, Tel., Fax., E-mail, Web-Site Address

LETTER TYPE: **PRECAUTIONARY NOTICE**
ADDRESSED TO: **ALL TRADES**
RE: **ASBESTOS DISCLOSURE NOTICE-1**

SCENARIO:

The following notice was placed on an asbestos abatement project.

COMPANY LETTERHEAD

(Recipient's Name)   (Date)
(Recipient's Title)
(Recipient's Contact Info.)

RE: (Project's Name and Tracking Number)

FOR YOUR SAFETY

DO NOT ENTER. THIS IS A HAZARDOUS CONSTRUCTION SITE.
TOXIC AND PATHOGENIC MATERIAL IS PRESENT.

This jobsite is restricted to authorized personnel only. Safety on the construction site is mandatory. The superintendent has been appointed to assure and enforce the safety program and safety meetings regarding all precautionary methods for this job.

If you have any questions regarding this notice, or need to talk to the general contractor or superintendent, please call the following numbers within the hours of 6:00 AM and 4:00 PM.

(500) 455-6777
(500) 455-6778

Your cooperation is appreciated.

Company Name, Address, Tel., Fax., E-mail, Web-Site Address

LETTER TYPE: **PRECAUTIONARY NOTICE**
ADDRESSED TO: **ALL TRADES**
RE: **ASBESTOS DISCLOSURE NOTICE-2**

SCENARIO:

The following notice was placed on an asbestos abatement project.

COMPANY LETTERHEAD

(Recipient's Name)   (Date)
(Recipient's Title)
(Recipient's Contact Info.)

RE: (Project's Name and Tracking Number)

FOR YOUR SAFETY

**THE FOLLOWING LIST REPRESENTS THE ASBESTOS-CONTAINING
PRODUCTS KNOWN TO BE PRESENT AT THIS JOBSITE:**

CEILINGS, WALLS AND INSULATION

Sprayed-on insulation on ceilings and walls
Fireproof wallboard
Insulation between plaster
Electrical insulation
Textured paint

ROOFING AND SIDING

Shingles
Sheets
Roofing felts

Please take all precautionary measures, as instructed by the superintendent, when working in close proximity to any of the products described above. All demolition and abatement should be handled by the asbestos-certified team only. Any violations will result in a reprimand.

Thank you for your cooperation.

Company Name, Address, Tel., Fax., E-mail, Web-Site Address

LETTER TYPE:	APOLOGETIC
ADDRESSED TO:	**COMPETITION**
RE:	**CONFLICT OF INTEREST**

SCENARIO:

Contractor A discovered that a new local contractor (B) has copied their marketing approach on registered or patented material. Contractor B gets a courtesy notice and decides to take action before it becomes a legal issue.

COMPANY LETTERHEAD

(Recipient's Name)   (Date)
(Recipient's Title)
(Recipient's Contact Info.)

RE: (Project's Name and Tracking Number)

Dear (Recipient's Name),

Please accept my sincere apology for conflicting with your business. I appreciate your courtesy letter.

At the time of the conflict, it did not seem to us that our marketing campaign mimicked your approach. It was only when your marketing director brought this matter to our attention that we realized your package appeared very similar to what we proposed.

Originally, I did not see any fault with our approach. However, I now realize that we need to change our strategies in order to avoid a conflict with you. I failed to research the competition more thoroughly.

Please be assured that I will take any necessary steps to correct this conflict by reworking our proposal approach. Please don't hesitate to contact me if you have any suggestions or concerns.

Respectfully,

(Name)
(Title)

Company Name, Address, Tel., Fax., E-mail, Web-Site Address

LETTER TYPE: CONGRATULATORY
ADDRESSED TO: SISTER COMPANY
RE: CONGRATULATORY LETTER, MARKETING

SCENARIO:

The CEO of a construction company wants to congratulate a sister company and offer his support. He also wants to explore the possibility of participating in parts of the job.

COMPANY LETTERHEAD

(Recipient's Name)   (Date)
(Recipient's Title)
(Recipient's Contact Info.)

RE: (Project's Name and Tracking Number)

Dear (Recipient's name),

We just heard the good news! I wanted to congratulate you for being the successful bidder on the new shopping mall project. We knew from the beginning that your experience in retail development would give you an edge on this bid.

In the past, we have had the opportunity and privilege to collaborate with your company on projects that were mutually beneficial. We would be pleased to support you with any aspect of the construction. Please let me know if there is anything we can do to assist you with this major undertaking.

Congratulations!

Best regards,

(Name)
(Title)

Company Name, Address, Tel., Fax., E-mail, Web-Site Address

LETTER TYPE: **RESPONDING TO A CONCERN**
ADDRESSED TO: **GENERAL CONTRACTOR**
RE: **DELAMINATION-1**

SCENARIO:

The general contractor for a new retail project is concerned about the effects of concrete delamination as seen in other projects executed under very similar climatic conditions in the area. The GC informs the subcontractor of this concern, and the concrete sub responds.

REFERENCE NOTE:
Use- BNi-W Form 225 as a Notice of Request for Technical Instructions
OR
Use- BNi-W Form 302 as a Notice of Request for Bid if soliciting sub-bids
OR
Use- BNi-W Form 273 for a general Proposal describing the scope of work.
OR
Use- BNi-W Form 113 as a Bid Confirmation describing the scope of work.

COMPANY LETTERHEAD

(Recipient's Name)   (Date)
(Recipient's Title)
(Recipient's Contact Info.)

RE: (Project's Name and Tracking Number)

Dear (Recipient's name),

I have read your e-mail regarding the delamination problem that you faced on other projects that share very similar pouring conditions. We are aware of this common problem and would like to offer some suggestions to assure compliance with the warranty of the lightweight slabs. We would strongly recommend that all lightweight concrete work for this project be finished with walk-behind power machines only. We will avoid the use of riding machines with pan floats.

On previous job surveys, we were informed that almost all delaminating cases were a result of not using the proper machines and finish processes, as opposed to the common misunderstanding that delamination is an aggregate problem. However, there are three factors that we should be aware of:

1. weather

2. mixture proportions

3. placement

We would strongly recommend scheduling the pour under the best controlled humidity and exposure.

I hope this offers you some direction or assistance. Please don't hesitate to call me if you need any additional information.

Sincerely,

(Name)
(Title)

Company Name, Address, Tel., Fax., E-mail, Web-Site Address

LETTER TYPE:
ADDRESSED TO: GENERAL CONTRACTOR
RE: DELAMINATION-2

SCENARIO:

The general contractor of the previous letter is still concerned and wants to make certain that other factors (outside finishing) will not impact the situation at hand. The subcontractor provides additional information on admixtures and air content in the slabs.

REFERENCE NOTE:
Use- BNi-W Form 225 as a Notice of Request for Technical Instructions
OR
Use- BNi-W Form 302 as a Notice of Request for Bid if soliciting sub-bids
OR
Use- BNi-W Form 273 for a general proposal describing the scope of work.
OR
Use- BNi-W Form 113 as a Bid Confirmation describing the scope of work.

COMPANY LETTERHEAD

(Recipient's Name)   (Date)
(Recipient's Title)
(Recipient's Contact Info.)

RE: (Project's Name and Tracking Number)

Dear (Recipient's name),

In regards to our current situation, I will do my best to explain the solution to your concerns. Even though I am not an engineer, I can rely on my twenty years of experience.

The traditional admixtures have chemicals that reduce the surface tension of water. You can think of these chemicals as a type of synthetic detergent that enhances the concrete by promoting bubbling. We need to use the right amount of air content to assure resistance to freezing and thawing. For your project, we will use a new polymer-based admixture that will assure the right bubbling consistency we need. We will re-temper the concrete if additional mixing is required.

We will be extremely careful when balancing the mixture content to compensate for humidity fluctuations. As explained before, we will use walk-behind power machines only.

I hope this eases your concerns. Feel free to contact me.

Respectfully,

(Name)
(Title)

Company Name, Address, Tel., Fax., E-mail, Web-Site Address

LETTER TYPE: NOTICE TO PERFORM
ADDRESSED TO: MECHANICAL SUBCONTRACTOR
RE: COMMISSIONING-1

SCENARIO:

A general contractor has to provide the fundamental commissioning requirements of the building's energy systems, as requested by the architect and US Green Building Council. The certified commissioning authority needs to be an independent mechanical sub. After obtaining bid quotes for this specific task, the general contractor wants to make sure the following items will be covered.

COMPANY LETTERHEAD

(Recipient's Name)   (Date)
(Recipient's Title)
(Recipient's Contact Info.)

RE: (Project's Name and Tracking Number)

Dear (Recipient's name),

We have reviewed your bid for the above referenced project. At this point, I would like to describe in more detail what we expect if the owner agrees with your quote and contract agreement. The following action items require your involvement:

1. You will be required to verify the building's related mechanical systems as they are installed, calibrated, and tested.

2. You will document the project requirements developed by the design team. You will make sure that these documents are complete and clear.

3. You will assist in the development and implementation of the overall plan.

4. You will be required to complete a summary commissioning report.

For a more precise scope of work, you will evaluate the heating, ventilating and air conditioning, and refrigeration (HVAC) systems. You will also cover any passive systems used in the project. The mechanical engineer will provide all necessary back-up information, such as cut-sheets and reading tables.

As the general contractor for this project, I would be interested in obtaining a certain level of training from you in regards to start-up balancing, testing, troubleshooting, and maintenance procedures. This training can be coordinated with our LEED AP.

Please let me know if you have any questions or concerns. Thank you for your time and interest in this project's commissioning bid.

Respectfully,

(Name)
(Title)

Company Name, Address, Tel., Fax., E-mail, Web-Site Address

LETTER TYPE: NOTICE TO PERFORM
ADDRESSED TO: MECHANICAL SUBCONTRACTOR
RE: COMMISSIONING-2

SCENARIO:

The general contractor of the previous project received a response from the mechanical subcontractor regarding the energy commissioning performance on his project. Apparently, the sub wants to deal directly with the client in order to avoid any type of mediatory intervention through the contractor. Even though the contractor respects this decision, he needs to establish a certain level of control on this job by writing the following letter.

COMPANY LETTERHEAD

(Recipient's Name)   (Date)
(Recipient's Title)
(Recipient's Contact Info.)

RE: (Project's Name and Tracking Number)

Dear (Recipient's name),

I appreciate your response and respect your right to negotiate directly with the client. However, we are under an enhanced commissioning agreement with the client, architect, and USGBC.

Under this enhanced commissioning process, the architect or contractor of the project also has a right to contact and contract with the commissioning authority. For the time being, it will be a good idea to stay in contact until our start-up meeting next week. At that point, you and the client can determine how you would like to proceed.

We are here to work collaboratively and we both look to serve the client. Please feel free to contact me if you have any questions prior to our meeting.

Respectfully,

(Name)
(Title)

Company Name, Address, Tel., Fax., E-mail, Web-Site Address

LETTER TYPE: **SCOPE VERIFICATION**
ADDRESSED TO: **SUBCONTRACTOR**
RE: **FIRE PROTECTION SYSTEM INSPECTION-1**

SCENARIO:

A general contractor needs to schedule the final testing and inspection for all the fire suppression systems in a building addition project.

COMPANY LETTERHEAD

(Recipient's Name)   (Date)
(Recipient's Title)
(Recipient's Contact Info.)

RE: (Project's Name and Tracking Number)

Dear (Recipient's name),

Per our conversation on (date), we are confirming that upon substantial project completion scheduled for (date), you will perform the routine inspection and testing required and included in the subcontract.

We must make sure that your inspection will also include foam-based fire suppression, and that the total number of fire suppression systems will be simultaneously tested. I would like to schedule the inspection so that my staff may temporarily vacate the building. The superintendent and myself will be present at the time of testing.

Respectfully,

(Name)
(Title)

Company Name, Address, Tel., Fax., E-mail, Web-Site Address

LETTER TYPE: SCOPE VERIFICATION
ADDRESSED TO: SUBCONTRACTOR
RE: FIRE PROTECTION SYSTEM INSPECTION-2

SCENARIO:

The general contractor of the previous letter receives a response to execute the fire suppression system testing. However, the inspector asks the contractor to prepare the jobsite by clearing all interior equipment and finishes that could be damaged as a result of the testing. The contractor responds.

COMPANY LETTERHEAD

(Recipient's Name)   (Date)
(Recipient's Title)
(Recipient's Contact Info.)

RE: (Project's Name and Tracking Number)

Dear (Recipient's name),

    Thank you for your response and commitment regarding the above referenced project. I have reviewed your list of prep-requirements, and agree to follow your instructions by adhering to established protocol activities prior to testing:

1. We will protect all switch gear equipment with impermeable blankets tightly secured at the top and sides of each cabinet. Since the cabinets are installed and mounted on concrete maintenance curbs, there will be no need to temporarily raise them.

2. We will locate a temporary sump-pump in areas prone to flooding, such as the battery room and engine rooms. Both rooms follow article (number) and are protected via containment with 4" concrete barriers.

3. We will use sandbags to divert any water runoff to the closest drains.

    We look forward to meeting with you on (date).

Respectfully,

(Name)
(Title)

Company Name, Address, Tel., Fax., E-mail, Web-Site Address

LETTER TYPE: **METHODOLOGY REQUEST**
ADDRESSED TO: **CONCRETE SUBCONTRACTOR**
RE: **SLAB JOINT REPAIR**

SCENARIO:

A general contractor is releasing a *48-Hour to Correct Deficiency in Workmanship* for the loose slab joints at a project. This problem is significant due to thermal expansion resulting from extreme climate changes. The contractor wants to make sure the best method is utilized to resolve the issue. Along with the attached notice, the contractor requires a specific repair methodology described as follows:

REFERENCE NOTE:
Use-BNi-W Form 218 as the 48-Hour Notice to Correct Deficiency in Workmanship.

COMPANY LETTERHEAD

(Recipient's Name)   (Date)
(Recipient's Title)
(Recipient's Contact Info.)

RE: (Project's Name and Tracking Number)

Dear (Recipient's name),

As previously brought to your attention, we are enclosing with this letter a *48-Hour Notice to Correct Deficiency in Workmanship* on the above referenced project. This notice is intended to target and resolve the loose slab joints in the lobby entryway and all perimeter walkways.

Please be aware that we will no longer accept traditional repair methodologies such as the use of retrofit dowels, joint replacements, semi-rigid joint fillers, or sub-slab grout injection. Please use the new standard method - the SD7 aluminum cylinder. The mechanical force achieved through the screw-and-wedge mechanism effectively locks the cylinder against each side of the concrete slabs. The internal spring will allow for 8,000 lbs. of force. As you are probably aware, this method is a permanent fix which offers a better warranty. Per our contract, please respond soon and before the 48-hour expiration.

Respectfully,

(Name)
(Title)

Company Name, Address, Tel., Fax., E-mail, Web-Site Address

LETTER TYPE: **INSTRUCTION**
ADDRESSED TO: **SUPERINTENDENT**
RE: **FAULTY ELECTRICAL INFRASTRUCTURE**

SCENARIO:

An electrical outage at a warehouse project forced the superintendent to stop work. The project is a structural retrofit of a URM (unreinforced masonry building). The existing electrical infrastructure dates back to the early 1920's, and was not part of the contract. Even though the contract does not cover electrical work, this may be a good opportunity to address this issue with the client as a possible addendum. The contractor needs to provide immediate instructions to the superintendent before addressing the client.

COMPANY LETTERHEAD

(Recipient's Name)   (Date)
(Recipient's Title)
(Recipient's Contact Info.)

RE: (Project's Name and Tracking Number)

Dear (Recipient's name),

I have been made aware of the electrical failure at the warehouse project. I will take this issue to the client and the architect in an attempt to evaluate the existing electrical infrastructure for repairs or replacements. I have already instructed our electrical sub to inspect the problem, and provide a comprehensive report with the following options:

Option 1:

Comprehensive list of electrical failures.

Estimate of repairs and schedule.

Option 2:

New electrical infrastructure list of components and unit pricing.

Labor estimate and schedule.

We expect to have this report no later than (date). Upon formal evaluation and cost assessment, the client and architect will have to make a determination so that we may re-define the scope of work. Regarding your request for troubleshooting, at this point it is very difficult to offer you proper assessment without inspecting the junction boxes and connections. It may be a result of loose connections that need tightening. If the existing wiring had ceramic caps, that may be the cause of the problem. If there is water intrusion at the plenum, it may have affected the conduit running on the uni-strut, which usually causes rusted connections. However, these are just possible troubleshooting ideas that have to be properly evaluated by the electrician.

Sincerely,

(Name)
(Title)

Company Name, Address, Tel., Fax., E-mail, Web-Site Address

LETTER TYPE:	**INSTRUCTION**
ADDRESSED TO:	**ARCHITECT**
CC:	**CLIENT**
RE:	**FAULTY CONNECTION-2**

SCENARIO:

The contractor for the warehouse project of the previous letter writes and e-mails the client and architect in order to set up a meeting to discuss the electrical outage problem and possible addendum.

REFERENCE NOTE:
Use-BNi-W Form 203 for Speed Memos when communicating with other firms and organizations.
OR
Use-BNi-W Form 203A for internal Speed Memos if you need a record reminder of the meeting.
OR
Use-BNi-W Form 212 for final Meeting Minutes Reporting.

REFERENCE NOTE:
Use- BNi-W Form 320 as a Notice of Directive or Communication if you need a written record of the architect, engineer or owner's order approval.
OR
Use- BNi-W Form 321 as a Notice of Reserving Impact Costs when additional costs and expenses on the remaining work arise.
OR
Use- BNi-W Form 103 as an Extra Work Order when extra work not covered by the original bid and contract is called for and should be executed.

COMPANY LETTERHEAD

(Recipient's Name)   (Date)
(Recipient's Title)
(Recipient's Contact Info.)

RE: (Project's Name and Tracking Number)

Dear (Recipient's name),

The intent of this message is to inform you that we had an electrical outage at the warehouse project. Even though our scope of work is the structural retrofit, we cannot continue overloading the existing electrical system with our tools and equipment. As you may already know, the electrical infrastructure for the building dates back to the late 1920's, and is not in compliance with today's code and load requirements. I would strongly recommend that you evaluate the existing system in order to determine an effective solution. The electrician for this job will inspect the problem and provide the following options:

Option 1:

Comprehensive list of electrical failures.

Estimate of repairs and schedule.

Option 2:

New electrical infrastructure list of components and unit pricing.

Labor estimate and schedule.

At this point, I would like to schedule a meeting at the jobsite with you, the architect, the electrical engineer, and the electrician to resolve this matter. Please consider the following date as a tentative schedule for this meeting, (date) at (time). Please let me know if this schedule will work.

Thank you for your time.

Respectfully,

(Name)
(Title)

Company Name, Address, Tel., Fax., E-mail, Web-Site Address

LETTER TYPE: **JUSTIFICATION**
ADDRESSED TO: **GENERAL CONTRACTOR**
CC: **SUBCONTRACTOR ANSWERING A COMPLAINT**
RE:

SCENARIO:

An electrical subcontractor received a written complaint from a general contractor regarding a poor installation. Since the subcontract with the electrician was handled directly by the client, a copy of this letter was sent to the client. The sub immediately takes action by responding to the complaint.

COMPANY LETTERHEAD

(Recipient's Name)   (Date)
(Recipient's Title)
(Recipient's Contact Info.)

RE: (Project's Name and Tracking Number)

Dear (Recipient's name),

I received your complaint regarding the conduit installation at the above referenced project. The job was done properly by identifying all incoming cables. Your sub was present when we identified the cable sheathing with markers. We marked all cables exiting the panel board as "origin-up stream." Subsequently, we marked the cables going to the outlets as "running line," and the following sequence outlet cables as "running-downstream."

I am confident that the installation was done correctly. However, we will go back and re-inspect our work. If there is something we neglected, we will fix it at our expense. Please let me know when we could schedule this inspection.

Respectfully,

(Name)
(Title)

CC (client)

Company Name, Address, Tel., Fax., E-mail, Web-Site Address

LETTER TYPE: INSTRUCTIONS
ADDRESSED TO: FRAMER/PLUMMER
CC: COORDINATION OF PIPING
RE:

SCENARIO:

A general contractor is coordinating instructions between his framer and the plumbing sub. He writes the following letter or e-mail.

COMPANY LETTERHEAD

(Recipient's Name)   (Date)
(Recipient's Title)
(Recipient's Contact Info.)

RE: (Project's Name and Tracking Number)

Dear (Recipient's name),

According to (subcontractor A), we need to drill fifteen bearing studs in order to run the plumbing for the above referenced job. The bearing studs have been designated and marked with a cross. The penetrations should not exceed the following criteria:

- Hole diameter on 2x4 bearing stud – 1 3/8" and 1/8" notch depth.
- Hole diameter on 2x6 bearing stud – 2 3/16" and 1 3/8" notch depth.

If the pipe's diameter exceeds the dimensions above, please notify me immediately so that we can determine if we need to use lumber reinforcement. Please make sure these requirements are met. The holes will be measured and inspected on (date).

Feel free to call me with questions or concerns.

Sincerely,

(Name)
(Title)

Company Name, Address, Tel., Fax., E-mail, Web-Site Address

LETTER TYPE: INSTRUCTIONS
ADDRESSED TO: GENERAL CONTRACTOR
RE: SITEWORK REVIEW

SCENARIO:

The owner of a construction company specializing in major site development projects has instructed his GC to perform a plan review before the excavation work starts. The project represents many drainage challenges that need to be addressed.

COMPANY LETTERHEAD

(Recipient's Name)   (Date)
(Recipient's Title)
(Recipient's Contact Info.)

RE: (Project's Name and Tracking Number)

Dear (Recipient's name),

Per our pre-construction meeting on (date), please familiarize yourself with the grading plan and sitework requirements. Make sure you understand where all surface site drainage is directed. In general, all site drainage will be directed to underground storm drain systems which will carry it offsite. Please inform the architect if there are areas that drain directly onto adjacent sites, if catch basins located in low points would back storm water into a structure, or if the catch basin becomes blocked.

Please review the project for catch basins and manholes, and submit catalog cuts to the architect if the structures are precast. Verify the spot elevations of tops and bottoms of drainage structures, and flow lines of storm drain pipes against your layout survey stakes. Review safety shoring requirements for excavations over five feet in depth.

In your overall project assessment, please also review the plans for possible conflicts between underground storm drains and other utility lines. Where conflicts are spotted, mark them on a set of drawings, and create a comprehensive list before excavation begins. Perform this review immediately, and consider a target meeting (date) to review your findings. Keep in mind, this job is very challenging and demands our full attention. Please contact me if you encounter any difficulties on this task.

Sincerely,

(Name)
(Title)

Company Name, Address, Tel., Fax., E-mail, Web-Site Address

Chapter Three
Industry Support Letters

This chapter helps the contractor establish a good record system of communication between various members of the construction industry, such as product manufacturers, union officials, insurance providers, etc. The tone of these letters is professional and cooperative. This is an excellent resource to assist in record-keeping, policy up-dates, evidence of cost increase, etc. These letters are extremely effective when used in combination with the forms listed below the *Letter Scenarios.*

LETTER TYPE:	**INFORMATION REQUEST**
ADDRESSED TO:	**SUPPLIER**
RE:	**REQUEST FOR PRODUCT APPROVAL**

SCENARIO:

A general contractor is trying to save on the cost of a metal frame building by implementing a new technology that promises to replace the use of assembled plywood shear walls.

REFERENCE NOTE:
Use-BNi-W Form 319 as a Notice of Request for Price
                                        OR
Use-BNi-W Form 206 as a Contract Invoice

COMPANY LETTERHEAD

(Recipient's Name)   (Date)
(Recipient's Title)
(Recipient's Contact Info.)

RE: (Project's Name and Tracking Number)

Dear (Recipient's Name),

We had the opportunity to discuss the development of your new product, the "Metal Wall System" on (date). We understand that the two year effort of testing the lateral bracing has proven to be a success.

At this point, we would like to inquire about the restrictions placed on your system regarding the allowable number of supported stories, as well as the ICC or approval numbers we need to use this product in the field.

I appreciate your time and assistance. Please call me if you have any questions regarding this request.

Best Regards,

(Name)
(Title)

Company Name, Address, Tel., Fax., E-mail, Web-Site Address

LETTER TYPE: DELIVERY COMPLAINT
ADDRESSED TO: PRODUCT SUPPLIER
RE: SHIPMENT DELAY

SCENARIO:

A general contractor placed a materials order for a tightly scheduled development project. The materials were not received by the promised delivery date, and the contractor informs the supplier.

COMPANY LETTERHEAD

(Recipient's Name)   (Date)
(Recipient's Title)
(Recipient's Contact Info.)

RE: (Project's Name and Tracking Number)

Dear (Recipient's name),

Due to the tardiness of the delivery of the (product) we ordered from you, we are now facing severe complications on the jobsite. We allowed you a (time allowed) to deliver the special order for the (product). As was explained to you, this was the maximum time permitted under our project schedule.

Unfortunately, (time period) have passed and we still don't have delivery. This delay is already causing us costly repercussions. At this time, we feel we are entitled to reimbursement for the costs associated with this delay. Please let me hear from you so that we can plan a resolution.

Respectfully,

(Name)
(Title)

Company Name, Address, Tel., Fax., E-mail, Web-Site Address

LETTER TYPE: **PRODUCT CANCELLATION**
ADDRESSED TO: **DISTRIBUTOR**
RE: **STOP PURCHASE ORDER**

SCENARIO:

The housing market has bottomed, and developers and clients have postponed or cancelled construction for the next six months. The general contractor has to act immediately by canceling the lumber he ordered in advance.

COMPANY LETTERHEAD

(Recipient's Name)   (Date)
(Recipient's Title)
(Recipient's Contact Info.)

RE: (Project's Name and Tracking Number)

Dear (Recipient's Name),

As you may already know, last year's construction market slowdown still continues. In the average adjusted annual projection, total construction has declined 8% from the previous month. This fact, along with overstocked housing units, has caused our clients to freeze development.

At this point, we are asking you to cancel the rest of the lumber orders assigned and approved for the next six months. We will be in constant communication with you, and the rest of our partner suppliers, hoping that a market recovery starts.

Please do not hesitate to contact me if you anticipate any problems regarding this cancellation. Thank you for your understanding.

Best regards,

(Name)
(Title)

Company Name, Address, Tel., Fax., E-mail, Web-Site Address

LETTER TYPE: CREDIT LIMIT INCREASE REQUEST
ADDRESSED TO: MANUFACTURER
RE: INCREASING CREDIT

SCENARIO:

A contractor has doubled his inventory. As a consequence, he needs to rapidly increase his company's credit limit with a local materials supplier.

COMPANY LETTERHEAD

(Recipient's Name)   (Date)
(Recipient's Title)
(Recipient's Contact Info.)

RE: (Project's Name and Tracking Number)

Dear (Recipient's Name),

Thank you for making credit available to us through your materials warehouse.

We are requesting that the credit limit initially established be increased due to our rapid growth. We are confident that our punctuality and payment history speaks for itself. Please let us know if a significant credit increase can be negotiated.

Please contact me at your earliest convenience.

Thank you for your time and consideration.

Cordially,

(Name)
(Title)

Company Name, Address, Tel., Fax., E-mail, Web-Site Address

LETTER TYPE: ENVIRONMENTAL REQUEST
ADDRESSED TO: PRODUCT SUPPLIER
RE: ENVIRONMENTAL PRODUCT CONSTRAINTS

SCENARIO:

A general contractor needs to make sure that the adhesives and sealants used for the construction of a new LEED-based school comply with established limits.

COMPANY LETTERHEAD

(Recipient's Name)   (Date)
(Recipient's Title)
(Recipient's Contact Info.)

RE: (Project's Name and Tracking Number)

Dear (Recipient's Name),

Before setting the product orders for the flooring adhesives and sealants, please confirm that your products conform to the following VOC Limit criteria:

Product Type	VOC Limit
Carpet Adhesive	50
Wood Floor Adhesive	50
Rubber Floor Adhesive	100
General Purpose Sealer	250
Membrane Roof	300
Roadway	250

Please let me know if your product line is consistent with the information provided above, and I will be happy to start processing the order. Please contact me at your earliest convenience. Thank you for your assistance.

(Name)
(Title)

Company Name, Address, Tel., Fax., E-mail, Web-Site Address

LETTER TYPE: REQUEST
ADDRESSED TO: PRODUCT SUPPLIER
RE: COATING SOLVENTS

SCENARIO:

A contractor writes a letter requesting clarification of the solvent type used for the coating he has selected to apply to a historic renovation project.

COMPANY LETTERHEAD

(Recipient's Name)   (Date)
(Recipient's Title)
(Recipient's Contact Info.)

RE: (Project's Name and Tracking Number)

Dear (Recipient's Name),

We are interested in using your coatings on our historic renovation project. This is due to the fact that your product closely matches the original varnish.

As you are aware, environmental constraints have resulted in more stringent requirements on solvents. Your product catalog states that your solvents are water-borne. Please explain if they are also water-based. We understand there is a clear distinction between water-borne and water-based solvents. As we understand, water-borne solvents can be chemically altered to receive water, whereas, the water-borne solvent is not truly a water-based product. For our project, we are strictly required to use water-based products only.

Thanks for your assistance.

Respectfully,

(Name)
(Title)

Company Name, Address, Tel., Fax., E-mail, Web-Site Address

LETTER TYPE: **INSURANCE INFORMATION REQUEST**
ADDRESSED TO: **INSURANCE PROVIDER**
RE: **POLICY REQUEST**

SCENARIO:

A contractor heard that there has been a decrease in insurance policy costs, and wants to improve his company's position.

COMPANY LETTERHEAD

(Recipient's Name)   (Date)
(Recipient's Title)
(Recipient's Contact Info.)

RE: (Project's Name and Tracking Number)

Dear (Recipient's Name),

I had the pleasure of meeting with you (date) to discuss the coverage of our current policy. At that time, we explained our operations with you in order for you to effectively communicate with the underwriters.

We are interested in a multi-year general liability policy to lock in current pricing and improve our coverage. Please let me know if there is any type of specialized product to cover the property loss exposure on any specific project.

I look forward to the list of policy options you have to offer.

Thank you for the excellent customer service and prompt response.

Best Regards,

(Name)
(Title)

Company Name, Address, Tel., Fax., E-mail, Web-Site Address

LETTER TYPE: **WARNING RELEASE**
ADDRESSED TO: **INSURANCE PROVIDER**
RE: **SUBROGATION WAIVER**

SCENARIO:

A project was destroyed by a fire. The investigation proved that the fire was caused by the contractor's welder. After paying for the damages, the client's builders risk insurer contacted the contractor in order to obtain his commercial general liability insurance information. The contractor knows this will mean that a claim will be placed against him through his insurance provider. He immediately reviews his contract and realizes that he is protected under the subrogation waiver clause. The contractor encloses a copy of the contract with the following letter.

COMPANY LETTERHEAD

(Recipient's Name)   (Date)
(Recipient's Title)
(Recipient's Contact Info.)

RE: (Project's Name and Tracking Number)

Dear (Recipient's Name),

    This letter is in regard to the fire that destroyed the above referenced project. The client's insurance provider paid for all damages. However, we understand that they will attempt to claim through you due to the fact that the fire was caused by our welder.

    Please be aware that we utilized the AIA contract document (number). This contract contains a *waiver of subrogation.* As you know, this clause protects us and our commercial general liability insurance provider from any liability due to a fire. The client's insurance company needs to see the waiver of subrogation clause in order to stop any claim against us.

    We are enclosing the contract for your records and use. Please call me if you need more information.

Respectfully,

Best Regards,

(Name)
(Title)

Company Name, Address, Tel., Fax., E-mail, Web-Site Address

LETTER TYPE: **INFORMATION REQUEST**
ADDRESSED TO: **INSURANCE PROVIDER**
RE: **INSURANCE COVERAGE**

SCENARIO:

A contractor received a judgment against him on a project that overexposed his liability. He is urgently requesting his GCL insurance agent to act on his behalf by describing the extent of his coverage against judgments and settlements.

COMPANY LETTERHEAD

(Recipient's Name)   (Date)
(Recipient's Title)
(Recipient's Contact Info.)

RE: (Project's Name and Tracking Number)

Dear (Recipient's Name),

    This letter is in reference to claim #(number). This claim explains my situation in full detail.

    It is urgent that I verify the coverage of my policy regarding a settlement offer on my behalf. I will need legal representation, and my insurance policy provides for the expense of a lawyer.

    Please let me know if you will appoint an attorney or if I should obtain my own.

    This request is time-sensitive. I appreciate your prompt response. Thank you.

Respectfully,

(Name)
(Title)

Company Name, Address, Tel., Fax., E-mail, Web-Site Address

LETTER TYPE: **INFORMATION REQUEST**
ADDRESSED TO: **INSURANCE PROVIDER**
RE: **WORKERS' COMP REQUEST**

SCENARIO:

A contractor is struggling to keep up with high workers' compensation insurance costs. He is requesting that his insurance agent provide him alternative or additional policy options.

COMPANY LETTERHEAD

(Recipient's Name)   (Date)
(Recipient's Title)
(Recipient's Contact Info.)

RE: (Project's Name and Tracking Number)

Dear (Recipient's Name),

    This letter is a request for assistance regarding our current workers' compensation insurance package. We understand that the recently enacted reforms, capped benefits, and improved medical claim reports, have lowered the rates by as much as 15%.

    Until now, we have utilized your standard product for our workers' compensation insurance. However, the costs of some of these have become unaffordable in some cases. We are now looking to you to provide partially self-insured packages, even though we may assume higher deductibles and risks as a result. Please provide us with some options regarding alternative policies, or let us know if the enacted reforms may help us qualify for better rates.

Respectfully,

(Name)
(Title)

Company Name, Address, Tel., Fax., E-mail, Web-Site Address

LETTER TYPE: RESTRICTIVE
ADDRESSED TO: UNION WORKERS
RE: DUAL GATE NOTICE (CONTROLLED ACCESS TO
                                JOBSITE)

SCENARIO:

After a union dispute, a general contractor is advised by his attorney to implement a "dual" or "reserve" gate system that will allow for controlled jobsite access. This letter is released as a notice to all union employees:

REFERENCE NOTE:
Use-BNi-W Form 216 if you want to use the actual Dual Gate Notice for Union Employees Form instead of the letter format

COMPANY LETTERHEAD

(Recipient's Name)   (Date)
(Recipient's Title)
(Recipient's Contact Info.)

RE: (Project's Name and Tracking Number)

Dear (Recipient's Name),

Please be aware that as the general contractor for this job, I must comply with all OSHA safety regulations. If your union believes any information or safety practice followed at this jobsite to be incorrect, please contact the National Labor Board with your concern. Due to the current labor dispute, at this time we are controlling access to and from the jobsite. A "dual" or "reserve" gate system has been lawfully established to permit you to freely enter, exit, and work on this project. You should know your rights under the dual or reserve gate system. You have the RIGHT to:

1. Be dispatched on a daily basis to this job;

2. Freely enter and exit the gate reserved for the use of your employer;

3. Enter and exit through a picketed gate if the picket line is illegal;

4. Work under a lawful, properly maintained dual gate system freely and without illegal threats of loss of pension, dental, medical, fringe benefits, work, "blackballing", or any other type of threatened penalty;

5. Work under a lawful, properly maintained dual gate system without any type of illegal threat of union internal discipline, such as suspension, expulsion or loss of trust fund credits.

You have a legal right and duty to enter and exit through the gate reserved for your employer. You also may lawfully enter and exit through a picketed gate if the picket line is illegal. The general contractor has a legal right to require your employer to man this job. Should your employer be unable to man this job on a day-to-day basis, your employer may be replaced on this job and backcharged for all damages caused by incomplete work. In return, your employer has the legal right to terminate or suspend any employee refusing to man this job as long as the gate reserved for your employer is not being lawfully picketed. If picketing on the gate reserved for your employer is illegal, you have the right to cross that illegal picket line and work. You may not be lawfully disciplined by the union for doing so. Please keep in mind we are enforcing these regulations for everyone's safety. Please don't hesitate to call me if you have any difficulties following these rules.

Thank you for your cooperation.

(Name)
(Title)

Company Name, Address, Tel., Fax., E-mail, Web-Site Address

LETTER TYPE: **ENVIRONMENTAL RESPONSE**
ADDRESSED TO: **UNION**
RE: **JOBSITE AIR QUALITY**

SCENARIO:

A general contractor received a warning regarding the lack of protection from air-borne pollutants in his remodel job. This directly affects his workers' pulmonary health. The contractor takes immediate corrective measures and responds to the union's claim.

COMPANY LETTERHEAD

(Recipient's Name)   (Date)
(Recipient's Title)
(Recipient's Contact Info.)

RE: (Project's Name and Tracking Number)

Dear (Recipient's Name),

Thank you for your concern regarding the health of my field staff. We have developed and implemented an indoor air quality management plan for the construction phases of our projects. With this plan, we will sequence the installation of materials to avoid contamination of absorptive materials such as insulation, carpeting, gypsum board, and ceiling tiles. We will also avoid using permanently installed air handlers for temporary heating or cooling during our remodeling jobs. Whenever possible, we will properly ventilate the work area. We will also require and enforce the use of filtered masks when air pollutants may be present.

Upon completion of the jobs we will "flush out" by supplying a total air volume of (amount) cu. ft. as recommended by our mechanical engineer. We will also conduct air testing after construction ends and prior to occupancy, using the testing protocols consistent with the United States Environmental Protection Agency Compendium of Methods for the Determination of Air Pollutants in Indoor Air. Please let me know if you have any suggestions for protecting the well-being of my employees.

Once again, thank you for bringing this issue to my attention.

Sincerely,

(Name)
(Title)

Company Name, Address, Tel., Fax., E-mail, Web-Site Address

Chapter Four
Letters to Personnel

This chapter covers the diverse aspects of daily internal communication. The tone of these letters varies from supportive to disciplinary. The employer needs to demonstrate a level of consistent professionalism in order to lead by example, and should be backed up by the right communication tools. These letters are extremely effective when used in combination with the forms listed below the *Letter Scenarios*.

LETTER TYPE:	**NOTICE**
ADDRESSED TO:	**CONSTRUCTION MANAGERS**
RE:	**MECHANICS' LIEN**

SCENARIO:

Internal memo released to the management staff of a medium-size construction company.

REFERENCE NOTE:
Use-BNi-W Form 104-A for the Notice to Owner Regarding Mechanics' Lien Law and Contractor's License Law.
OR
Use-BNi-W Form 106 for the Mechanic's Lien Filing Form.

COMPANY LETTERHEAD

(Recipient's Name)   (Date)
(Recipient's Title)
(Recipient's Contact Info.)

RE: (Project's Name and Tracking Number)

Attention all Management Staff:

It has been a common request from management to define and describe company procedures regarding a mechanics' lien filing.

We follow the guidelines provided by law: a mechanic's lien must be recorded within 90 days after the completion of the work as a whole, unless the owner records a notice of completion. If a notice of completion is recorded, the mechanics' lien must be recorded within 30 days thereafter unless the claimant is a general contractor or specialty contractor who contracted directly with the owner. In that case, the mechanics' lien must be recorded within 60 days after the notice of completion was recorded.

A mechanics' lien expires unless a foreclosure suit is filed within 90 days after the lien was recorded. The Mechanics' Lien Law is frequently amended.

Please see me if you have any questions about the procedure. If I don't have the answer, we will consult with our attorney. Thank you for your cooperation.

Best Regards,

(Name)
(Title)

Company Name, Address, Tel., Fax., E-mail, Web-Site Address

LETTER TYPE: **EMPLOYMENT NOTICE STATUS**
ADDRESSED TO: **JOB CANDIDATE**
RE: **NEGATIVE DRUG TEST RESULTS**

SCENARIO:

A job applicant whose drug test came back with a positive result was sent the following letter.

COMPANY LETTERHEAD

(Recipient's Name)   (Date)
(Recipient's Title)
(Recipient's Contact Info.)

RE: (Project's Name and Tracking Number)

Dear (Recipient's Name),

This letter is to inform you that the drug testing for the "Special Risk Position" you applied for came back positive. We maintain a drug-free workplace in accordance with the standards and procedures established under the articles of the Workers' Compensation Law.

This information is strictly confidential and will be given only to a medical review officer to determine if prescription or nonprescription medications were present at the time of testing. At this time, you have the legal right to contest or explain the result to the medical review officer within 5 working days after receiving written notification of the test result. Please find attached a list of medications, which may alter or affect a drug test provided by the Agency for Health Care Administration.

By law, we are required to disclose the following information:

"An employer may not discharge, discipline, refuse to hire, discriminate against, or request or require rehabilitation of an employee or job applicant on the sole basis of a positive test result that has not been verified by a confirmation test and by a medical review officer."

For more information, please contact our Human Resources department between the hours of (time).

Sincerely,

(Name)
(Title)

Company Name, Address, Tel., Fax., E-mail, Web-Site Address

LETTER TYPE: **CONGRATULATORY**
ADDRESSED TO: **PERSONNEL**
RE: **OUTSTANDING JOB PERFORMANCE-1**

SCENARIO:

Regardless of obstacles in the construction of an office building, the superintendent in charge demonstrated a high level of commitment and hard work. His efforts resulted in an outstanding job delivered and finished ahead of schedule.

COMPANY LETTERHEAD

(Recipient's Name)   (Date)
(Recipient's Title)
(Recipient's Contact Info.)

RE: (Project's Name and Tracking Number)

Dear (Recipient's Name),

I was delighted to hear the latest news regarding your outstanding performance on the construction of our new building at (address). Your general contractor proudly described your level of commitment and attention to detail. Your efforts are highly appreciated. This remarkable performance demonstrates your leadership ability, as well as your organizational skills. I want to express my deep appreciation for your efforts.

It is because of valuable people like you that our organization is ranked among the highest in the industry. I hope you find the enclosed bonus as a token of our appreciation.

Keep up the good work!

Best Regards,

Sincerely,

(Name)
(Title)

Company Name, Address, Tel., Fax., E-mail, Web-Site Address

LETTER TYPE: **CONGRATULATORY**
ADDRESSED TO: **PERSONNEL**
RE: **OUTSTANDING JOB PERFORMANCE-2**

SCENARIO:

After congratulating the superintendent in the previous letter, the owner of the company decides to congratulate the rest of the construction team that participated in the building project.

COMPANY LETTERHEAD

(Recipient's Name)   (Date)
(Recipient's Title)
(Recipient's Contact Info.)

RE: (Project's Name and Tracking Number)

Dear Team,

It is an honor for me to congratulate you on the successful completion of the new building at (address). I hope you look back on this great accomplishment and take it as an example of what you are able to accomplish as a team. I also want to take this opportunity to express my personal appreciation for your contribution. Enclosed is an invitation to a special event dinner in your honor.

I look forward to seeing you there to congratulate each and every one of you in person.

Best regards,

(Name)
(Title)

Company Name, Address, Tel., Fax., E-mail, Web-Site Address

LETTER TYPE: **CONGRATULATORY**
ADDRESSED TO: **PERSONNEL**
RE: **RETIREMENT CONGRATULATIONS**

SCENARIO:

The senior accountant for a construction company has retired, and the owner wants to recognize her efforts.

COMPANY LETTERHEAD

(Recipient's Name)   (Date)
(Recipient's Title)
(Recipient's Contact Info.)

Dear (Recipient's name),

Congratulations on your retirement! Thank you for helping to make this company so successful for the (number) of years you have been with us. Your accounting skills, knowledge, and expertise in the construction industry have distinguished you in our company.

We will be forever grateful for all your achievements and your pleasant personality, and we hope your years of retirement are filled with well-deserved relaxation and unrestricted time with your family and loved ones. You will be missed.

Again, thank you for all your efforts and for the wonderful years.

Congratulations!

Warmest Regards,

(Name)
(Title)

Company Name, Address, Tel., Fax., E-mail, Web-Site Address

LETTER TYPE:	NOTICE
ADDRESSED TO:	**PROJECT MANAGERS**
RE:	**EMPLOYMENT FREEZE**

SCENARIO:

The CEO of a construction company needs to cut costs during the decline of the housing market. He approaches his management staff with an order affecting new hires.

COMPANY LETTERHEAD

(Recipient's Name)   (Date)
(Recipient's Title)
(Recipient's Contact Info.)

RE: (Project's Name and Tracking Number)

To all Management:

As you may already know, there has been a dramatic recession in the housing market. We have reported a 60% decrease in our already shrinking margins. As a result, we will be forced to take some drastic measures in order to avoid further losses. Effective immediately, all hiring for framers and finishers will be suspended. This will also affect temporary and part time staff.

All field workers who leave the company during the next (number) months will not be replaced until further notice. All submitted applications will be kept on file for six months. Please be prepared to suggest ways to implement further cost and expense reductions on our current projects. There will be an upcoming meeting to discuss this issue. Thank you for bearing with us during this hardship.

Respectfully,

(Name)
(Title)

Company Name, Address, Tel., Fax., E-mail, Web-Site Address

LETTER TYPE: **REPRIMAND**
ADDRESSED TO: **FIELD EMPLOYEE**
RE: **TARDINESS**

SCENARIO:

A formwork worker repeatedly fails to arrive on time to the jobsites. The general contractor writes a reprimand.

COMPANY LETTERHEAD

(Recipient's Name)   (Date)
(Recipient's Title)
(Recipient's Contact Info.)

RE: (Project's Name and Tracking Number)

Dear (Name):

   Please be aware that your shift begins promptly at (time).   In fairness to your superintendent and team, it is critical that you arrive on time.

   When you were hired, we clearly explained how important punctuality is in this business. We work as a team, and we respond collectively to the task at hand. Without your timely participation, the team is missing a tool. Tardiness affects everyone. We need you to take this commitment seriously.

   I sincerely hope this will be the last time you need to be told about the importance of punctuality. Thank you for your cooperation.

Sincerely,

(Name)
(Title)

Company Name, Address, Tel., Fax., E-mail, Web-Site Address

LETTER TYPE: REPRIMAND
ADDRESSED TO: PERSONNEL
RE: EQUIPMENT

SCENARIO:

Some employees in the central office of a small construction company have used the computer to process and store personal information. This information was discovered by the owner of the company, and he responds with this reprimand.

COMPANY LETTERHEAD

(Recipient's Name)   (Date)
(Recipient's Title)
(Recipient's Contact Info.)

RE: (Project's Name and Tracking Number)

To All Staff,

   We ask all employees to refrain from using the company computers for anything other than business-related communication. Please be aware that all e-mails and on-line transactions are saved and archived on a continuous basis. The use of the internet is also monitored.

   We will not tolerate improper use of our equipment. We appreciate your cooperation in this matter.

(Name)
(Title)

Company Name, Address, Tel., Fax., E-mail, Web-Site Address

LETTER TYPE: **REPRIMAND**
ADDRESSED TO: **PERSONNEL**
RE: **POOR PERFORMANCE**

SCENARIO:

A welder is showing a decreasing level of quality in his work. This problem has intensified, and the general contractor decides to address it.

COMPANY LETTERHEAD

(Recipient's Name)   (Date)
(Recipient's Title)
(Recipient's Contact Info.)

RE: (Project's Name and Tracking Number)

Dear (Name),

Over the past (number) years, you have demonstrated the ability to be a skilled welder for our company. However, we have noticed that for the past two jobs, your quality of work has been drastically deteriorating.

As you realize, proper welding is critical in ensuring the structural integrity of the steel connections within the frame of a building. We encourage you to revisit your skills and make the effort required for this job before it becomes a liability issue.

Thank you for your cooperation.

Sincerely,

(Name)
(Title)

Company Name, Address, Tel., Fax., E-mail, Web-Site Address

LETTER TYPE: INTERVIEW INVITATION
ADDRESSED TO: PERSONNEL
RE: EMPLOYMENT INTERVIEW

SCENARIO:

A position for a field engineer opened, and after advertising for potential candidates and receiving resumes, the owner of the company decides to offer the job to a candidate he feels may be a good fit for the position.

COMPANY LETTERHEAD

(Recipient's Name)   (Date)
(Recipient's Title)
(Recipient's Contact Info.)

RE: (Project's Name and Tracking Number)

Dear (Name),

Thank you for applying for the position of field engineer.

We have reviewed your resume, and feel that your progressive experience on engineering projects shows the level of competence we are looking for in our employees.

At this time, we would like to invite you for an interview. Please call me at your earliest convenience to set up a date and time when you are available to discuss your potential with our organization.

I look forward to hearing from you.

Respectfully,

(Name)
(Title)

Company Name, Address, Tel., Fax., E-mail, Web-Site Address

LETTER TYPE: NOTICE
ADDRESSED TO: PERSONNEL
RE: CRAFT TRAINING

SCENARIO:

A construction company is struggling to obtain more skilled craftsmen for a rapidly growing market. The general contractor decides to invest in training for his installers.

COMPANY LETTERHEAD

(Recipient's Name)   (Date)
(Recipient's Title)
(Recipient's Contact Info.)

RE: (Project's Name and Tracking Number)

To all Installers:

After careful review of our current situation, we have decided to implement a technical training program.

It is imperative that we stay ahead of the high levels of technical proficiency required for our business.

The training will include a mandatory six-month technical apprenticeship program regulated by the FTI. We will schedule groups of three installers per each session. The selection will be based on the employee's current schedule.

Please be aware that a meeting will be scheduled shortly to inform you of further details.

Thank you very much for your cooperation.

Respectfully,

(Name)
(Title)

Company Name, Address, Tel., Fax., E-mail, Web-Site Address

LETTER TYPE: DISCIPLINARY
ADDRESSED TO: PERSONNEL
RE: INAPPROPRIATE BEHAVIOR

SCENARIO:

A masonry worker repeatedly harassed a female employee with inappropriate flirting and sexual innuendo. The contractor offers an apology and fires the worker. He decides to take further action by writing the following memo to the rest of his staff:

COMPANY LETTERHEAD

(Recipient's Name)   (Date)
(Recipient's Title)
(Recipient's Contact Info.)

RE: (Project's Name and Tracking Number)

Dear Valued Employees,

This memo is to bring to your attention the issue of improper behavior. An employee has been terminated this morning due to such behavior. We are taking additional steps in order to avoid a reoccurrence of this situation. We have scheduled an emergency meeting to discuss this issue with the staff and subcontractors. We have also formulated the following list:

1. All employees will conduct themselves in a respectful and professional manner
2. Any inappropriate behavior will be cause for immediate termination
3. The use of inappropriate language will not be tolerated
4. Any inappropriate behavior should be reported immediately to the superintendent
5. If a client has a concern, let them know that you will promptly address the issue with the superintendent
6. Respect the GC and superintendent's authority at the jobsite and follow their instructions.

You will be notified as to the time and date of this meeting. Thank you for your cooperation.

(Name)
(Title)

Company Name, Address, Tel., Fax., E-mail, Web-Site Address

LETTER TYPE: NOTICE TO TAKE ACTION
ADDRESSED TO: PERSONNEL (PROJECT MANAGER)
RE: NEW CONTRACT

SCENARIO:

A construction company has been selected to build a series of projects for a large organization. The contract requires a demonstration of building operations sequencing. The owner of the company alerts and prepares his management staff for the meeting.

COMPANY LETTERHEAD

(Recipient's Name)   (Date)
(Recipient's Title)
(Recipient's Contact Info.)

RE: (Project's Name and Tracking Number)

MEMO:

Our company has been chosen as the general contractor for 25 post office projects. We have been required under the contract to demonstrate our scheduling expertise.

It is critical that you are present for this important meeting, which will be held this coming (day) at (time) at (address). I look forward to seeing you at the meeting.

Thank you for your cooperation.

(Name)
(Title)

Company Name, Address, Tel., Fax., E-mail, Web-Site Address

LETTER TYPE:	**ACCIDENT ACKNOWLEDGEMENT**
ADDRESSED TO:	**PERSONNEL**
RE:	**WORK RELATED INJURY**

SCENARIO:

A construction worker has fallen from a scaffold. He is immediately taken to the hospital and admitted as an inpatient. The owner of the company is on vacation. He is notified via e-mail and responds with the following letter.

COMPANY LETTERHEAD

(Recipient's Name)   (Date)
(Recipient's Title)
(Recipient's Contact Info.)

RE: (Project's Name and Tracking Number)

Dear (Recipient's Name),

I was distressed to hear about your accident. I understand you were immediately attended to, and on your way to a recovery. Please be aware that your insurance matters will be handled by the company.

Don't worry about the job. The important thing for now is that you follow your doctor's orders. I will be in contact with (name) from our Human Resources department and will check on your progress. If you need anything, please contact (name), or feel free to e-mail me with any questions or concerns. I will be back in the office next (day). Take care.

Sincerely,

(Name)
(Title)

Company Name, Address, Tel., Fax., E-mail, Web-Site Address

LETTER TYPE: **PERFORMANCE REVIEW**
ADDRESSED TO: **PERSONNEL**
RE: **SALARY ADJUSTMENT**

SCENARIO:

A framer has demonstrated reliability and progress on the job. The owner of the company wants to recognize and promote this valuable employee.

COMPANY LETTERHEAD

(Recipient's Name)   (Date)
(Recipient's Title)
(Recipient's Contact Info.)

RE: (Project's Name and Tracking Number)

Dear (Recipient's Name),

   I enjoyed meeting with you and (name of supervisor) for your annual performance review. When obstacles were presented, you took proactive actions to resolve them. We appreciate your responsible attitude. Thank you for your contribution in making our projects flow smoothly.

   At this time, we would like to promote you to a new position in the company. Your new title is "Field Supervisor." Your salary has also been reviewed and reflects a (amount) per hour increase.

   Congratulations and keep up the good work!

Best Regards,

(Name)
(Title)

Company Name, Address, Tel., Fax., E-mail, Web-Site Address

LETTER TYPE:	<mark>**NOTICE TO FILE**</mark>
ADDRESSED TO:	**ACCOUNTS RECEIVABLE**
RE:	**RELEASE OF PAYMENT BOND**

SCENARIO:

The owner of a medium-size construction company is instructing his staff on how to file a legal document.

REFERENCE NOTE:
Use-BNi-W Form 334 for a Release of Labor and Material Bond

COMPANY LETTERHEAD

(Recipient's Name)   (Date)
(Recipient's Title)
(Recipient's Contact Info.)

RE: (Project's Name and Tracking Number)

Dear (Recipient's Name),

Please find enclosed the "Release of Payment Bond" form regarding the office building at (address). This form should be used when a payment bond has been posted to make certain that the payment of money due, owed, and unpaid after the completion of the bonded work on the project will be available. The purpose of the release form is to clear the outstanding bond. When completed, this form should be sent to the bonding company to verify that the bonding company no longer has liability under the bond issued.

Any legal document, contract, and state-specific references should be reviewed by our attorney for possible revisions to ensure compliance with local laws. This particular form has already been reviewed by our legal department.

Please remember that the form must be signed in the presence of a notary.

Consider this instruction a high priority. If you have any questions, please contact me. Thank you.

Sincerely,

(Name)
(Title)

Company Name, Address, Tel., Fax., E-mail, Web-Site Address

LETTER TYPE: **REMINDER NOTICE**

ADDRESSED TO: **FINISHER**

RE: **SANDBLASTING**

SCENARIO:

The contractor for a major commercial project is instructing his staff to avoid sandblasting the existing brick walls.

COMPANY LETTERHEAD

(Recipient's Name)   (Date)
(Recipient's Title)
(Recipient's Contact Info.)

RE: (Project's Name and Tracking Number)

Dear (Recipient's Name),

   I am sending you this message as a reminder that you should only sandblast the existing stone veneer at the shopping center project. <u>Do not sandblast the brick walls</u>. These walls will be refinished with a new protective coating. Sandblasting removes the hard protective surface on manufactured products.

   As you know, most of the walls on this project are manufactured masonry. If cleaning masonry is absolutely necessary, be sure that you first test a hidden or obscure spot. This way, we can avoid unacceptable results.

   If you have any problems during the cleaning process, please contact me immediately. Thank you for your cooperation.

Sincerely,

(Name)
(Title)

Company Name, Address, Tel., Fax., E-mail, Web-Site Address

LETTER TYPE: **NOTICE**
ADDRESSED TO: **FIELD EMPLOYEES**
RE: **WASTE REDUCTION POLICIES**

SCENARIO:

In an attempt to help the environment, the state has imposed strict recycling programs though local HAZMAT authorities regarding waste diversion. The general contractor for a major airport remodel has been targeted this year for these new regulations. The contractor has reviewed the new regulations, and wants to make sure he follows the requirements in order to do his part for the environment and avoid getting penalized. He establishes a training program in which recycling containers are placed on the jobsite. Regardless of this fact, the demolition and clean-up crew fails to follow instructions. To correct this problem, the contractor writes the following notice.

COMPANY LETTERHEAD

(Recipient's Name)   (Date)
(Recipient's Title)
(Recipient's Contact Info.)

RE: (Project's Name and Tracking Number)

To All Demolition and Clean-Up Crews:

It has come to my attention that most of the clean-up team is not following our **Waste Reduction Policy** as established in last month's training.

We have invested time and resources to make sure we follow these mandatory **HAZMAT** regulations. If we fail to comply, we will get sanctioned. As a reminder, we have designated recycling areas at the north side of the jobsite. Four main disposal containers have been placed:

1. The green container is for paper, cardboard balers, packaging, TYVEK, construction paper, and backing. Do not discard adhesives in this container.

2. The blue container is for metal scraps, aluminum cans, sheet metal flashing cuts, etc.

3. The yellow container is strictly for drywall and cement products. Do not use as a concrete wash area.

4. The red container is strictly for hazardous waste. This container is a health hazard, and is only for use by authorized personnel certified for hazmat abatement.

The measures described above will be strictly enforced. Anyone not following these regulations will be suspended from the jobsite. Your cooperation with this policy is appreciated.

Sincerely,

(Name)
(Title)

Company Name, Address, Tel., Fax., E-mail, Web-Site Address

LETTER TYPE:	NOTICE
ADDRESSED TO:	PARTNER
RE:	LOSING PROFIT

SCENARIO:

A construction company is losing money on a project. The general contractor has reviewed his budget, payment schedules, cost records, change orders, and any accounting documents in the attempt to understand the problem. He realizes that his partner negotiated the contract on a "lump-sum" basis. This offered the client an advantage in negotiating the job. Unfortunately, the lump-sum percentage was too low for the scale of the project. The general contractor needs to address this issue with his partner.

COMPANY LETTERHEAD

(Recipient's Name)   (Date)
(Recipient's Title)
(Recipient's Contact Info.)

RE: (Project's Name and Tracking Number)

Dear (Name):

I would like to follow up on a conversation we had this morning regarding our unprofitable bank project. We agreed that the problem resulted from using a lump-sum contract. We need to use a cost-plus contract from now on. With this contract type, we have better control of risk by allowing additional costs outside of the original bid. Then, as long as we follow through on each project diligently, we are sure to make a profit.

We need to be careful not to jeopardize our profits by lack of record control or by lack of attention to our project. I am asking you to pay more attention to the accounting and bookkeeping procedures that go into this type of contract. Keep in mind that with smaller jobs, we need to charge a higher percentage in order to keep our profit and overhead controlled. Please feel free to ask me for guidance if you need to determine the percentage for each job.

Thank you for your cooperation.

Sincerely,

(Name)
(Title)

Company Name, Address, Tel., Fax., E-mail, Web-Site Address

LETTER TYPE:	INSTRUCTIONS
ADDRESSED TO:	**PROJECT MANAGER**
RE:	**DELEGATING RESPONSIBILITY-1**
	(PRE-CONSTRUCTION MEETING.)

SCENARIO:

The owner of a medium size construction company will be traveling on business and needs to contact his project manager regarding a pre-construction meeting on an awarded project.

COMPANY LETTERHEAD

(Recipient's Name)   (Date)
(Recipient's Title)
(Recipient's Contact Info.)

RE: (Project's Name and Tracking Number)

Dear (Name):

I will be traveling on business for the new resort/casino project from (date) to (date). Regarding the water station project, please make sure you conduct a pre-construction meeting before (date). I need to record this meeting to start billing for the job. You will be responsible for reporting back to me with the meeting minutes. Please use an action item version when reporting your minutes.

This format easily allows me to set action items into potential cost items that we can then pass on to accounting for billing purposes. Let's be proactive and think about more efficient ways to handle the upcoming meeting. Make sure you discuss the following ten issues:

1. Duties and responsibilities of owner, builder, and subs.
2. Review contract agreement, if necessary.
3. Review insurance requirements.
4. Establish a payment schedule.
5. Review the project schedule.
6. Review inspection schedule and requirements.
7. Review security procedures.
8. Review change order procedure.
9. Review clean-up procedures.
10. Establish responsibility for obtaining permits.

You can always reach me via e-mail. I will check my messages every night.

Sincerely,

(Name)
(Title)

Company Name, Address, Tel., Fax., E-mail, Web-Site Address

LETTER TYPE: **INSTRUCTIONS**
ADDRESSED TO: **PROJECT MANAGER**
RE: **DELEGATING RESPONSIBILITY-2**
                              (CLOSE-OUT DOCUMENTS)

SCENARIO:

The owner of the construction company on the previous letter gives closing instructions to another project manager regarding a recently completed project.

COMPANY LETTERHEAD

(Recipient's Name)   (Date)
(Recipient's Title)
(Recipient's Contact Info.)

RE: (Project's Name and Tracking Number)

Dear (Name):

I will be traveling on business for the new resort/casino project from (date) to (date). Regarding the manufacturing plant project, please provide all close-out documentation to the owner and architect by (date).

Please make sure you include the following documents:

1. Executed punch list   (Get signature from architect)
2. Waiver of lien certificates   (Obtain from subs)
3. Equipment warranties   (Obtain from manufacturers)
4. Equipment operating instructions
5. Certificates of inspection
6. Occupancy certificate
7. Contractor affidavit

You can always reach me via e-mail. I will check my messages.

Sincerely,

(Name)
(Title)

Company Name, Address, Tel., Fax., E-mail, Web-Site Address

LETTER TYPE: INSTRUCTIONS
ADDRESSED TO: SUPERINTENDENT
RE: DELEGATING RESPONSIBILITY
                      FALL PROTECTION PLAN

SCENARIO:

The owner of a construction company in charge of a multi-story office building reviewed the safety standards and methodology for the *Fall Protection Plan* guidelines established by the *OSHA* inspector. The owner of the company designates a person to be the security monitor.

COMPANY LETTERHEAD

(Recipient's Name)   (Date)
(Recipient's Title)
(Recipient's Contact Info.)

RE: (Project's Name and Tracking Number)

Dear (Name):

Please make sure we comply with the protective measures reviewed by the safety inspector. We need to work closely with him in order to assure adequate fall protection by identifying the locations where conventional control lines can't be provided.

I have designated you as the safety monitor due to your experience with construction safety programs. One of your duties will be to document and list all potential hazards. The items on your list will be discussed in the safety pre-construction training scheduled for (date).

If you have any questions, please call me at the office. Thank you for your assistance.

Sincerely,

(Name)
(Title)

Company Name, Address, Tel., Fax., E-mail, Web-Site Address

LETTER TYPE: **REPRIMAND**
ADDRESSED TO: **SUPERINTENDENT**
RE: **FIRE EXTINGUISHERS**

SCENARIO:

The general contractor for a major communications network station is informed of a fire resulting from inadequate fire protection. This negligence resulted in loss of equipment. The contractor needs to reprimand the superintendent in charge of the job.

COMPANY LETTERHEAD

(Recipient's Name)   (Date)
(Recipient's Title)
(Recipient's Contact Info.)

RE: (Project's Name and Tracking Number)

Dear (Name):

It was brought to my attention that your failure to follow proper procedure led to the loss of equipment on the jobsite. Keep in mind that this project is sensitive due to the equipment infrastructure within its core. There are generator rooms, battery rooms, and computer network systems linked to a series of servers throughout the building. The architect specified dry chemical fire extinguishers only. When the fire started in the server room, you carelessly used a foam extinguisher. This caused severe damage to the equipment, and we are now forced to replace it at our own expense.

Keep in mind that foam is a conductor and should never by used on a type "C" fire. Make sure you follow the architect's specifications and our own construction safety plans. In case of fire, activate the alarm immediately, utilize the correct type of extinguisher, use the equipment correctly and without delay. Make sure the fire is completely extinguished, recheck equipment, and run a test to assess any damages.

Let this be the last incident resulting from lack of attention. Thank you for your cooperation.

(Name)
(Title)

Company Name, Address, Tel., Fax., E-mail, Web-Site Address

LETTER TYPE: REQUEST
ADDRESSED TO: SUPERINTENDENT
RE: OVERTIME

SCENARIO:

A shopping mall project represents a major challenge for a medium size construction company. The owner of the company is under extreme pressure to finish the project on schedule. The contractor needs to ask for more hours by requesting overtime from all project managers.

COMPANY LETTERHEAD

(Recipient's Name)   (Date)
(Recipient's Title)
(Recipient's Contact Info.)

RE: (Project's Name and Tracking Number)

All Management Staff:

I want to thank all of you for your positive attitude and flexibility in regards to this project. As mentioned earlier, it is essential that we complete this project on schedule. In order to achieve this aggressive goal, we need everyone's participation by working overtime for the next month. Please conduct a team meeting as soon as possible to coordinate your schedules. You will report with the schedule requirements and hourly needs in our next management meeting.

Please be aware that a meal period of thirty minutes every four hours will be provided and paid for by the company. You and your team will be paid according to the overtime rate established by law and company policy. In addition to this, we will make sure everyone gets rewarded with a bonus upon project close-out.

I appreciate your assistance and cooperation.

Sincerely,

(Name)
(Title)

Company Name, Address, Tel., Fax., E-mail, Web-Site Address

LETTER TYPE:	REQUEST
ADDRESSED TO:	ACCOUNTING
RE:	PREVAILING WAGE-1

SCENARIO:

The contractor for a residential development project needs to follow the prevailing wage law enforced for public improvements. Even though this is a private community, its infrastructure needs are tied to the city's utility infrastructure and is therefore governed by prevailing wages.

COMPANY LETTERHEAD

(Recipient's Name)   (Date)
(Recipient's Title)
(Recipient's Contact Info.)

RE: (Project's Name and Tracking Number)

All Management Staff:

We understand that all private projects involving public improvements such as roads, sewer, and water lines will be subject to prevailing wages. We also know that most private developments we have worked on in the past were exempt from the prevailing wage requirements. It was brought to my attention by our attorney that certain amendments to (state section number) will probably be enacted.

I need to make a final assessment with our attorney before authorizing you to proceed with payroll. Please expect my decision no later than (date).

Sincerely,

(Name)
(Title)

Company Name, Address, Tel., Fax., E-mail, Web-Site Address

LETTER TYPE: REQUEST
ADDRESSED TO: ACCOUNTING
RE: PREVAILING WAGE-2

SCENARIO:

This is a follow-up for the previous letter.

COMPANY LETTERHEAD

(Recipient's Name)   (Date)
(Recipient's Title)
(Recipient's Contact Info.)

RE: (Project's Name and Tracking Number)

All Management Staff:

This letter is a follow-up on the prevailing wage involvement for this project.

Please find an enclosed copy of the non-performance report that would exclude us from prevailing wage requirements regarding site grading and trenching only. For the utility infrastructure and connections to the municipal grid, please complete the statement of Employer Payments (form number), and submit at the time we accept the subcontract bid.

In order to avoid a penalty, keep all prevailing wage payment records in a separate file. Before making final payments, please request an affidavit from the subcontractors stating that they have paid their employees the specified prevailing wage rate.

If you have any questions feel free to contact me.

Thank you for your assistance.

Sincerely,

(Name)
(Title)

Company Name, Address, Tel., Fax., E-mail, Web-Site Address

LETTER TYPE:	REPRIMAND
ADDRESSED TO:	**SUPERINTENDENT**
RE:	**INCOMPLETE RECORD KEEPING**

SCENARIO:

A general contractor is trying to collect information from an inspection report as well as his superintendent's written response. After failing to find it in the job-file system, the contractor decides to address this issue with the superintendent.

COMPANY LETTERHEAD

(Recipient's Name)   (Date)
(Recipient's Title)
(Recipient's Contact Info.)

RE: (Project's Name and Tracking Number)

Dear (Name):

It has come to my attention that the file system for the above job is missing the written records we urgently need to justify cost and process.

Please provide all written correspondence in a timely manner, and archive in its job-file folder. Release this communication to all involved parties and provide me with a copy. Proper and timely distribution of correspondence is extremely important. The construction inspector should receive copies of all material concerning the contract documents.

Although some inspectors may take notes on scraps of paper, it is critical to establish a written record as soon as possible. Remember that verbal information usually results in confusion. Make sure you keep an accurate written record in your laptop. I will be asking you to submit your correspondence with the inspector and anyone else involved in the field.

Thank you for your cooperation.

Sincerely,

(Name)
(Title)

Company Name, Address, Tel., Fax., E-mail, Web-Site Address

LETTER TYPE: INSTRUCTIONS
ADDRESSED TO: LABOR STAFF
RE: INSTRUCTIONS

SCENARIO:

A general contractor is scheduling the slab pouring for a project, and needs to pull laborers from other trades in order to sustain the project's momentum. A framer, with no concrete experience, is brought in to help. The general contractor instructs him.

COMPANY LETTERHEAD

(Recipient's Name)   (Date)
(Recipient's Title)
(Recipient's Contact Info.)

RE: (Project's Name and Tracking Number)

Dear (Name):

In regard to our conversation on (date) about the concrete slabs to be poured at the jobsite on (day), I am sending you this e-mail as a reminder of what should be done.

Since you are new to the process, the sub will walk you through before pouring. Keep in mind that the footings are poured first, and then vibrated. Do not attempt to screed the job on your own. We usually screed with a crew of no less than three men. Two men pull the screed while a third one fills low spots with a rake.

Please leave finishing to someone with experience. Bull-floating requires a mastering technique that you will learn in time. The use of the bull-floating is not as easy as it looks. Please concentrate on spreading the concrete and vibrating it. The superintendent will coordinate your team. Again, please meet with the superintendent on (date/time).

Sincerely,

(Name)
(Title)

Company Name, Address, Tel., Fax., E-mail, Web-Site Address

LETTER TYPE:	**INCONVENIENCE JUSTIFICATION**
ADDRESSED TO:	**NEIGHBOR**
RE:	**ANSWERING NEIGHBORHOOD COMPLAINT**

SCENARIO:

The general contractor of a retail project received a complaint from one of the neighbors. He decides to respond in order to avoid any conflicts.

COMPANY LETTERHEAD

(Recipient's Name)   (Date)
(Recipient's Title)
(Recipient's Contact Info.)

RE: (Project's Name and Tracking Number)

Dear (Recipient's Name),

Please accept our apologies regarding the noise levels at the above referenced project. We are making every effort to minimize the nuisance and discomfort to our neighbors. However, any project of this type is bound to cause some level of inconvenience. I have instructed my field workers to minimize noise levels as much as possible.

We sincerely feel that you will ultimately be happy with the project, since it will provide you with some great new services. The grand opening has been scheduled for (date). We would be delighted if you can join us for the celebration. We will send you an invitation when the time approaches.

As a token of appreciation, we would like to offer you a complimentary gift package, courtesy of the store owners.

Thank you for your patience and understanding.

Best Regards,

(Name)
(Title)

Company Name, Address, Tel., Fax., E-mail, Web-Site Address

INDUSTRY ORGANIZATIONS

In addition to the organizations and agencies listed in Part 2 of this book, many other professional and trade organizations and agencies listed below represent the members of the construction industry. Many of these groups publish documents and provide other information relating to design, materials, systems, specifications, and construction.

National Organizations

Acoustical Society of America, 120 Wall St, New York, NY 10005, (212) 248-0373.

Adhesive and Sealant Council, 1627 K St NW, STE 1000, Washington, DC 20006, (202) 452-1500 - FAX: (202) 452-1501.

Advisory Council on Historic Preservation, The Old Post Office Bldg, 1100 Pennsylvania Ave NW, STE 809, Washington, DC 20004, (202) 606-8503 - FAX: (202) 606-8647.

Air Conditioning and Refrigeration Institute, 4301 N Fairfax Dr, #425, Arlington, VA 22203, (703) 524-8800 - FAX: (703) 524-8800

Air Conditioning and Refrigeration Wholesalers, 1351 South Federal Highway, PO Box 640, Deerfield Beach, FL 33441.

Air Conditioning Contractors of America, 1228 17th St NW, Washington, DC 20036.

Air Pollution Control Association, PO Box 2861, Pittsburgh, PA 15230.

Allied Stone Industries, Carthage Marble Co, Carthage, MO 64836.

American Arbitration Association, 140 W. 51st St, New York, NY 10020, (212) 484-4006, (800) 778-7879.

American Association for Hospital Planning, Century Bldg., STE. 830, 2341 Jefferson Davis Hwy, Arlington, VA 22202.

American Association of Junior Colleges, National Center for Higher Education, One DuPont Circle NW, Washington, DC 20036.

American Association of Museums, 1575 I St NW, STE 400, Washington, DC 20005, (202) 289-1818 - FAX: (202) 289-6578.

American Association of School Administrators, 1801 N Moore St, Arlington, VA 22209, (703) 528-0700 - FAX: (703) 841-1543 (800) 771-1162.

American Concrete Institute, PO Box 9094, Farmington Hills, IL 48333.

American Concrete Pipe Association, 8300 Boone Blvd, #400, Vienna, VA 22182.

American Construction Inspectors Association, 2275 W. Lincoln Ave, STE. B, Anaheim, CA 92801.

American Forest Council, 1111 19th St NW, #800, Washington, DC 20036, (202) 463-2455.

American Forestry Association, 1319 18th St NW, Washington, DC 20036.

American Gas Association, 1515 Wilson Blvd, Arlington, VA 22209, (703) 841-8400.

American Hardware Manufacturers Assoc, 801 N Plaza Dr, Schaumberg, IL 60173, (312) 885-1025 - FAX: (847) 605-1093.

American Hospital Association, 840 N Lakeshore Dr, Chicago, IL 60611, (312) 280-6000.

American Institute of Architects, 1735 New York Ave NW, Washington. DC 20006, (202) 626-7300 (800) 242-3837.

American Institute of Kitchen Dealers, 124 Main St, Hackettstown, NJ 07840.

American Institute of Landscape Architects, 6810 N Second Pl, Phoenix, AZ 85012.

American Institute of Planners. 1313 E. 60th St, Chicago, IL 60637.

American Institute of Real Estate Appraisers of the National Assn. of Realtors, 875 N Michigan, STE 2400, Chicago, IL 60611 (312) 335-4100.

American Insurance Association, 1130 Connecticut Ave NW, #1000, Washington, DC 20036, (202) 828-7100 - FAX: (202) 293-1219.

American Iron and Steel Institute, 1101 17th St NW, Washington, DC 20005, (202) 452-7100.

American Library Association, 50 E. Huron St, Chicago, IL 60611, (312) 944-6780 (800) 545-2433.

American National Standards Institute, 11 W. 42nd St, 13th Floor, New York, NY 10036, (888) 267-4783 - FAX: (212) 398-0023.

American Plywood Association, 7011 S. 19th St, Tacoma, Washington 98466 (253) 565-6600.

American Public Power Association, 2301 M St NW, Washington, DC 20037, (202) 775-8300 FAX: (202) 467-2910.

American Public Works Association, 1313 E. 60th St, Chicago, IL 60637, (773) 667-2200 - FAX: (773) 667-2304.

American RD and Transportation Builders Association, ARBA Bldg, 1010 Massachusetts Ave, Washington, DC 20001, (202) 289-4434.

American Segmental Bridge Institute, 9201 N 25th Ave, STE 150B, Phoenix, AZ 85021, (602) 997-9964 - FAX: (602) 997-9965.

American Society of Civil Engineers, 1801 Alexander Bell Dr, Reston, VA 20190, (800) 548-2723.

American Society of Golf Course Architects, 221 N LaSalle St, Chicago, IL 60601, (312) 372-7090 - FAX: (312) 372-6160.

American Society of Heating, Refrigeration and Air Conditioning Engineers, Inc, 1791 Tullie Circle N.E, Atlanta, GA 30329, (404) 636-8400 - FAX: (404) 321-5478 (800) 527-4723.

American Society of Mechanical Engineers, United Engineering Center, 345 E. 47th St, New York, NY 10017 (212) 705-7722 (800) THE-ASME.

American Society of Planning Officials, 1313 E. 60$^{th\ St}$, Chicago, IL, 60637, (312) 947-2560.

American Society of Real Estate Counselors, 430 N Michigan Ave, Chicago, IL 60611, (312) 329-8431 - FAX: (312) 329-8881.

American Society for Testing & Materials, 100 Barr Harbor Dr. W, Conshocken, PA 19428 (610) 832-9585.

American Welding Society, Inc, PO Box 351040, 550 NW 42nd Ave, Miami, FL 33135.

American Wood Preservers Association, PO Box 286, Woodstock, MD 21663 (410) 465-3169.

Architectural Precast Association, 825 E. 64th St, Indianapolis, IN 46220, (317) 251-1214.

Asphalt Institute, Research Park Dr, Lexington, KY 40512 (606) 288-4961.

Association Builders and Contractors, Inc, 1300 N 17th St, Roslyn, VA 22209, (703) 812-2000.

Associated Equipment Distributors, 615 W. 22nd St, Oakbrook, IL 60523, (630) 574-0650.

Associated General Contractors of America, 1957 E St NW, Washington, DC 20006, (202) 393-2040.

Associated Specialty Contractor, Inc, 3 Bethesda Metro Center #1100, Bethesda, MD 20814, (301) 657-3110.

Association of American Universities, One DuPont Circle, Washington, DC 20036 (202) 387-3760.

Association of University Architects, c/o Forrest M. Kelly, Jr., Physical Planning Officer State University System of Florida Collins Bldg, Tallahassee, FL 32301.

Association of Wall and Ceiling Industries International, 1600 Cameron St, Alexander, VA 22314-2705, (703) 684-2924 FAX: (703) 684-2935.

Association of Women in Architecture, 7440 University Dr, Saint Louis, MO 63130 (314) 621-3484.

Better Heating-Cooling, 35 Russo Pl, Berkeley Heights, NJ, (908) 464-8200.

Builders Hardware Manufacturers Association, 355 Lexington Ave, 17th Fl, New York, NY 10017, (212) 661-4261 - FAX: (212) 370-9047.

BLDG Congress and Exchange, 2301 N Charles St, Baltimore, MD 21218.

BLDG Materials Research Institute, Inc, 15 E. 40th St, New York, NY 10017.

BLDG Research Institute, 2101 Constitution Ave NW, Washington, DC 20418.

BLDG Stone Institute, 420 Lexington Ave, New York, NY 10017, (212) 490-2530.

BLDG Systems Research Institute, 2101 Constitution Ave NW, Washington, DC 20418.

BLDG Thermal Envelope Coordinating Council, 101 15th St NW, STE. 700, Washington, DC 20005, (202) 347-5710.

California Association of Realtors, 525 S. Virgil Ave, Los Angeles, CA 90020, (213) 739-8200.

Ceilings and Interior Systems Contractors Association, 1500 Lincoln Hwy, STE 202, St. Charles, IL 60174 (630) 584-1919.

Cellular Concrete Association, 715 Boylston ST, Boston, MD 02116.

Ceramic Tile Distributors Association, 15 Salt Creek Lane, STE. 422 Hinsdale, IL 60521.

Ceramic Tile Institute, 700 N Virgil Ave, Los Angeles, CA, 90029.

Committee of Steel Pipe Producers American Iron And Steel Institute, 1000 16th St NW, Washington. DC 20036.

Concrete Reinforcing Steel Institute. 933 N Plluli Grove RD Schaumburg, IL. 60173-4758, (312) 517-1200 - FAX: (312) 517-1206.

Construction Financial Management Assn, 40 Brunswick Ave, Edison, NJ 08818.

Appendix-Industry Organizations

Construction Labor Research Council, 1730 M ST NW, STE 503, Washington, DC 20036.

Construction Specifications Institute, 601 Madison St, Alexandria, VA 22314, (800) 689-2900.

Construction Writers Association, PO Box 5586, Buffalo Grove, IL 60089 (847) 398-7756.

Contracting Plasterers Research Institute, 2101 Constitution Ave NW, Washington, DC 20418.

Copper Development Association, Inc, 260 Madison Ave - 16th Floor, New York, NY 10016 (212) 251-7200 (800) 232-3282.

Council of Educational Facility Planners, 29 W. Woodruff Ave, Columbus, OH 43210.

Council of Mechanical Specialty Contracting Industries, Inc, 7315 Wisconsin Ave, Washington, DC 20014.

Electrical Association, 140 S. Dearborn ST, Chicago, IL 60603.

Electric Power Research Institute, 2000 L St, STE 805 NW, Washington, DC 20036 (202) 872-9222 - FAX: (202) 293-2697.

The Energy Bureau, Inc, 41 E. 42nd St, New York, NY 10017.

Engineers Joint Council, 345 E. 47th St, New York, NY 10017.

Federal Housing Administration, Dept. of Housing and Urban Development, 451 7th St SW, Washington, DC 20410 (202) 708-2495 - FAX: (202) 708-2583.

Fine Hardwoods Association, 5603, West Raymond, STE 0, Indianapolis, IN 46241, (317) 873-8780.

Flexicore Manufacturers Association, PO Box 1807, Dayton, OH 45401, (937) 223-7420.

Food Facilities Consultants Society, 1800 Pickwick Ave, Glenview, IL 60025.

Forest Products Research Society, 2801 Marshall Ct, Madison, WI 53705, (608) 231-1361 (800) 354-7164.

Gardens For All, 180 Flynn Ave, Burlington, VT 05401.

Guild For Religious Architecture, 1913 Architects Bldg, Philadelphia, PA 19103.

Historic American BLDGs Survey, 801 19th St NW, Washington, DC 20006.

Illuminating Engineers Society, 120 Wall St, New York, 17th Floor, NY 10005 (212) 248-5000.

Information Bureau of Lath/Plaster/Drywall, 3127 Los Feliz Blvd, Los Angeles, CA 90039, (213) 663-2213.

Institute of Electrical and Electronic Engineers, 345 E. 47th St, New York, NY 10017, (212) 705-7900 (800) 678-4333.

Institute of Noise Control Engineering, PO Box 3206, Arlington Branch, Poughkeepsie, NY 12603.

Institute of Real Estate Management, 430 N Michigan Ave, Chicago, IL 60611 (312) 661-1930 (800) 837-0706 - FAX: (800) 338-4736.

International Association of Plumbing and Mechanical Officials, 2001 S. Walnut Dr, Walnut, CA 91789-2825.

International Conference of Building Officials, 14545 Leffingwell, Whittier, CA 90604 (562) 903-1478 - FAX: (561) 903-1480.

International Council of Shopping Centers, 665 Fifth Ave, New York, NY 10022, (212) 421-8181 - FAX: (212) 421-6464.

International Institute of Ammonia Refrigeration, 111 East Wacker Dr, Chicago, IL 60601, (312) 644-6610 - FAX: (312) 565-4658.

International Masonry Institute, 815 15th St NW, Washington, DC 20005, (202) 783-3908.

Inter-Society Color Council, Inc, Rensselaer Polytechnic Institute, Troy, NY 12181.

Landscape Architecture Foundation, 636 I ST NE, Washington, DC 20001 (202) 898-2444.

Mason Contractors Association of America, 1910 So. Highland Ave, STE 101, Lombard, IL 60148 (630) 705-4200 - FAX: (630) 705-4209.

Masonry Institute of America, 2550 Beverly Blvd, Los Angeles, CA 90057 (213) 388-0472 FAX: (213) 388-6958

Metal Buildings Manufacturers Association, c/o Thomas Assoc, 1300 Summer Ave, Cleveland, OH 44115 (216) 241-7333 - FAX: (216) 241-0105.

Model Codes Standardization Council, National Bureau of Standards, Washington, DC 20234.

Mortar Manufacturers Standards Association, 315 S. Hicks Rd, Palatine, IL 60067.

Mortgage Bankers Association of America, 1125 15th ST NW, Washington, DC 20005, (202) 861-6500.

National Asphalt Pavement Association, 5100 Forbes Blvd. Lanham MD 20706, (301) 731-4748.

National Association of Corrosion Engineers, 1440 South Creek Dr, Houston, TX 77084, (281) 492-0535.

National Association of Decorative Architectural Finishes, 112 N Alfred St, Alexandria, VA 22314.

National Association of Garage Door Manufacturers, 1300 Summer Ave, Cleveland OH 44115 (216) 241-7333.

National Association of Home Builders National Housing Center, 1201 15th St NW, Washington, DC 20005 (202) 822-0200 (800) 223-2665.

National Association of Home Builders, 1201 15th St NW, Washington, DC 20005, (202) 822-0200.

National Association of Housing and Redevelopment Officials, 630 Eye St NW, Washington, DC 20001 (202) 289-3500.

National Association of Realtors, 700 11th St NW, Washington, DC 20001 (202) 283-1043.

National Association of Store Fixture Manufacturers, 5975 W. Sunrise, Sunrise, FL 33312 (305) 587-9190.

National Board of Boiler and Pressure Vessel Inspectors. 1055 Crupper Ave, Columbus, OH 43229, (614) 888-8320.

National Institute of Standards and Technology, Fire and BLDG Research Labs, Gaithersburg, MD 20899.

National Concrete Masonry Association, 2302 Horse Pen Rd, Herndon, VA 22071 (703) 713-1900.

National Construction Association, 1730 M St NW, STE 503, Washington, DC 20036, (202) 466-8880.

National Corporation for Housing Partnership, 1133 15th St NW, Washington, DC 20005, (202) 216-2900.

National Crushed Stone Association, 1415 Elliott PI NW, Washington, DC 20007.

National Decorating Products Assn, 415 Ax Minister, St Louis, MO 63026.

National Electrical Contractors Association, Inc, 3 Betheseda, MD 20814.

National Fire Protection Association, 1 Batterymarch Park, Quincy, MA 02269 (800) 344-3555.

National Housing Conference, 815 15th ST NW, STE 711, Washington, DC 20005 (202) 393-5772.

National Institute of BLDG Sciences, 1201 L ST NW, #400, Washington, DC 20005, (202) 289-7800.

National Petroleum Council, 1625 K ST NW, STE. 600, Washington, DC 20006, (202) 393-6100.

National Precast Concrete Association, 10333 N Meridian St., STE 272, Indianapolis, IN 46290 (317) 571-9500.

National Ready Mixed Concrete Association, 900 Spring St, Silver Springs, MD 20910 (301) 587-1400 - FAX: (301) 585-4219.

National Roofing Contractors Association, 10255 W. Higgins Rd, STE 600, Rosemont, IL 60018 (708) 299-9070.

National Science Foundation, 4201 Wilson Blvd, Arlington, VA 22230 (703) 306-1070.

National Slag Association, 900 Spring St, Silver Springs, MD 20910 (301) 587-1400.

National Wood Flooring Association, 233 Old Meramec Stations Rd, Manchester, MO 63021 (314) 391-5161.

North American Wholesale Lumber Association, 3601 Algonquin Rd, STE 400, Rolling Meadows, IL 60008 (708) 870-7470.

Painting and Decorating Contractors of America, 3913 Old Lee Highway, STE. 33-B, Fairfax, VA 22030, (703) 359-0826 - FAX: (703) 359-2576 (800) 332-7322.

Plastering Information Bureau, 21243 Ventura Blvd, STE 115, Woodland Hills, CA 91364.

Plastic in Construction Council, 355 Lexington, New York, NY 10001.

Plumbing and Drainage Institute, PO Box 93, Indianapolis, IN 46206, (317) 251-5298.

Portland Cement Association, 5420 Old Orchard Rd, Skokie, IL 60077. (800) 868-6733.

Prestressed Concrete Institute, 175 W. Jackson Blvd, Chicago, IL 60604, (312) 786-0300.

Red Cedar Shingle and Handsplit Shake Bureau, 515 116th Ave. NE, STE. 275, Bellevue, WA 98004, (425) 453-1323.

Scaffold Industry Association, Inc, 14039 Sherman Way, Van Nuys, CA 91405, (818) 782-2012 - FAX: (818) 786-3027.

Scaffolding and Shoring Institute, c/o Thomas Associates, Inc, 1300 Summer, OH 44115, (216) 241-7333.

Screen Manufacturers Association, 2850 S. Ocean Blvd, No. 311, Palm Beach, FL 33480, (407) 533-0991.

Sealed Insulating Glass Manufacturers Association, 401 N Michigan Ave, Chicago, IL 60611, (312) 644-6610.

Sheet Metal and Air Conditioning Contractors National Assn, Inc, 4201 Lafayette Center Dr, Chantilly, VA 20151 (703) 803-2989.

Sheet Metal and Air Conditioning Contractors' National Association, Inc, 4201 Lafayette Center Dr, Chantilly, VA 20151 (703) 803-2980.

Society of the Plastic Industry, 1275 K St NW, Washington, DC 20005, (202) 371-5200.

Solar Energy Industries Association, 122 C St NW, 4th Floor, Washington, DC 20001, (202) 383-2600.

Southern Cypress Manufacturers Association, 400 Penn Center Blvd, Pittsburgh, PA 15235 (412) 829-0770.

Stained Glass Association of America, PO Box 22642, Kansas City, MO 64113 (816) 333-6690.

Steel Door Institute, 30200 Detroit Rd, Cleveland, OH 44145, (216) 226-0010.

Stucco Manufacturers Association, 507 Evergreen, Pacific Grove, CA 93950 (408) 649-3466.

Truss Plate Institute, 583 D'Onofrio Dr., STE 200, Madison, WI 53719 (608) 833-5900.

United Brotherhood of Carpenters and Joiners of America, 101 Constitution AVE NW, Washington, DC 20001, (202) 546-6206.

United States Conference of Mayors, 1620 Eye St NW, Washington, DC 20006, (202) 293-7330.

United States League of Savings Institutions, 111 E. Wacker Dr, Chicago, IL 60601, (312) 644-3100.

Urban Institute, 2100 M St NW, Washington, DC 20037, (202) 624-7062.

Vermiculite Association, 11 S. La Salle St, STE 1400, Chicago, IL 60603 (312) 201-0101.

Waferboard Assn, PO Box 724533, Atlanta, GA 30339.

Wallcovering Information Bureau, 66 Morris Ave, Springfield, NJ 07081.

Wallcovering Wholesalers Association, 401 N Michigan Ave, Chicago, IL 60611 (312) 245-1083.

Western Red Cedar Lumber Association, 1500 Yeon Bldg, Portland, OR 97204.

Western Wood Products Association, 522 SW 5th Ave, STE 500, Portland, OR 97204 (503) 224-3930.

Wood and Synthetic Flooring Institute. 1800 Pickwick AVE, Glenview, IL 60025.

Wood Truss Council of America, 1 WTCA Center, 6425 Normandy Ln, Madison, WI 53719 (608) 274-3329.

CALIFORNIA ORGANIZATIONS

Air Conditioning and Refrigeration Contractors Association of Southern California, 401 Shatto PI, Los Angeles, CA 90020, (213) 738-7238 (213) 738-5260.

American Public Works Association, Northern California Chapter

American Subcontractors Association, Los Angeles/Orange County Chapter, c/o Philip B. Greer, Atkinson, Andelson, et al, 13304 E. Alondra Blvd, STE 200, Cerritos, CA 90701.

Associated Builders and Contractors, Golden Gate Chapter, 11875 Dublin Blvd, STE 258 Dublin, CA 94568, (510) 829-9230

Associated General Contractors of California, East Bay District, 1390 Willow Pass RD, STE 1030, Concord, CA 94520 (510) 827-2422

Associated General Contractors of California, State Office, 3095 Beacon Blvd, West Sacramento, CA 95691, (916) 371-2422.

Associated Plumbing and Mechanical Contractors of Sacramento, 50 Fullerton CT, #100, Sacramento, CA 95825, (916) 452-4917 FAX: (916) 452-0532.

Associated Roofing Contractors of the Bay Area Counties, 8301 Edgewater Dr, Oakland, CA 94621, (510) 635-8800.

Associated Tile Contractors of Southern California, 2736 S. La Cienega Blvd, Los Angeles, CA 90034.

Builders Exchange of Alameda, 3055 Alvarado St, San Leandro, CA 94577, (510) 483-8880.

Builders Exchange of Contra Costa, 1900 Bates Ave, Suites E & F, Concord, CA 94520 (510) 685-8630.

Builders Exchange of Modesto, PO Box 4307, Modesto, CA 95352, (209) 522-9031.

Builders Exchange of Monterey Peninsula, 343 Ocean AVE, Monterey, CA 93940, (408) 373-3033 - FAX: (408) 373-8682.

Builders Exchange of Napa/Solano, 135 Camino Dorado, Napa, CA 94558, (707) 255-2515.

Builders Exchange of the Peninsula, 735 Industrial RD, San Carlos, CA 94070, (650) 591-4486 FAX: (650) 591-8108.

Builders Exchange of Salinas Valley, 590-A Brunched AVE, STE A, Salinas, CA 93901, (408) 758-1624 - FAX: (408) 758-6203.

Builders Exchange of San Francisco, 850 S. Van Ness AVE, San Francisco, CA 94110, (415) 282-8220.

Builders Exchange of Santa Clara, 400 Reed St, Santa Clara, CA 95050, (408) 727-4000.

Builders Exchange of Santa Cruz, 2555 So. Cal Dr, Santa Cruz, CA 95065 (408) 476-6349.

Builders Exchange of Stockton, 7500 N West Lane (plans only), Stockton, CA 95210, PO Box 8040 (letters only), Stockton CA 95208, (209) 478-1005 - FAX: (209) 478-2132.

California Association of Realtors, 525 So. Virgil AVE, Los Angeles, CA 90020, (213) 365-9256 - FAX: (213) 365-9256.

Appendix-Industry Organizations

California BLDG Industry Association, 1107 9ᵗʰ St, STE 1060, Sacramento, CA 95814, (916) 443-7933 - FAX: (916) 443-1960.

California Conference of Masonry Contractor Associations, 7844 Madison Ave, STE. 140, Fair Oaks, CA 95628.

California Contractors Association, 6055 E. Washington Blvd, STE 200, Los Angeles, CA 90040, (213) 726-3511 - FAX: (213) 726-2366.

California, Division of State Architecture.

California Landscape Contractors Association, 2021 N St #300, Sacramento, CA 95814, (916) 448-2522.

California OSHPD.

California Wall and Ceiling Contractors Association, 1111 Town and Country Rd. #45, Orange, CA 92668.

Ceramic Tile Institute of Northern California, 10408 Fair Oaks Blvd, Fair Oaks, CA 95628, (916) 965-8453 - FAX: (916) 965-8454.

Concrete Masonry Association of California and Nevada, 6060, Sunrise Vista Dr, Citrus Heights, CA 95610, (916) 722-1700.

Concrete Pumpers Association of Southern California, 1567 Colorado Blvd, Los Angeles, CA 90041, (213) 257-5266.

Construction Industry Research Board, 2511 Empire AVE, Burbank, CA 91504, (818) 8341-8210.

Contractors Bonding Association, 529 W. Imperial Way, STE 5, Los Angeles, CA 90044.

El Dorado Builders Exchange, 2808 Mallard LN #B, Placerville, CA 95667, (530) 622-8642.

Electric Power Research Institute, 3412 Hillview AVE, Palo Alto, CA 94304, (415) 855-2000.

Electric Contractors of California and Nevada, 7700 Edgewater Dr. #640, Oakland, CA. 94621.

Engineering Contractors Association, 8310 Florence AVE, Downey, CA 90240, (562) 861-0929.

Floor Covering Institute, 400 Reed St, Santa Clara, CA 95050, (408) 727-4320.

Fresno Builders Exchange, PO Box 111, Fresno, CA 3707, CA 95667, (209) 237-1831.

Independent Roofing Contractors of California, 3478 Buskirk AVE #1040, Pleasant Hill, CA 94523, (510) 939-3715.

Kern County Builders Exchange, 1121 Baker St, Bakersfield, CA 93305, (805) 324-5364.

Los Angeles County Painting and Decorating Contractors Association, Inc, 1106 Colorado Blvd, Los Angeles, CA 90041, (213) 258-8136 - FAX: (213) 258-2279.

Marin Builders Exchange, 110 Belvedere St, San Rafael, CA 94901, (415) 456-3222.

Masonry Contractors Association of Sacramento, 7844 Madison AVE, STE 140, Fair Oaks, CA 95628, (916) 966-7666.

Mechanical Contractors Legislative Council of California, 7 Crow Canyon CT, STE. 200, San Ramon, CA 94583.

Merced-Mariposa Builders Exchange, PO Box 761, Merced, CA 95341, (209) 722-3612.

Minority Contractors Association of Los Angeles, 3707 W. Jefferson, Los Angeles, CA 90016, (213) 737-7952.

National Association of Women in Construction of Los Angeles, PO Box 90935, Pasadena, CA 91109.

National Association of Women in Construction, 4865 Pasadena AVE, Sacramento, CA 95841, (916) 483-2724.

National Electrical Contractors Association, Los Angeles Chapter, 401 Shatto Pl, Los Angeles, CA 90020, (213) 487-7313 - FAX: (213) 388-5230.

North Coast Builders Exchange, 216 W. Perkins St, Ukiah, CA 954:82, (707) 462-9019.

North Coast Builders Exchange, 987 Airway Ct, Santa Rosa, CA 95403, (707) 542-9502.

Northern California Drywall Contractors Association, 12241 Saratoga-Sunnyvale RD, Saratoga, CA 95070, (408) 255-1544.

Northern California Engineering Contractors Association, 3354 Regional Prkwy, Santa Rosa, CA 95403, (707) 525-1910.

Pacific Coast Builders Conference, 605 Market St, San Francisco, CA 94105, (415) 821-3307.

Painting and Decorating Contractors of California, 3504 Walnut AVE, STE A, Carmichael, CA 95608, (916) 972-1055 - FAX: (916) 972-9831.

Painting and Decorating Contractors of Central Coast Counties, 4050 Ben Lomond Dr, Palo Alto, CA 94306, (650) 493-6200.

Painting and Decorating Contractors of Sacramento, 3913 Old Lee Highway, STE 33B, Fairfax, VA 22030 (800) 332-7322.

Peninsula Builders Exchange, 735 Industrial Rd, San Carlos, CA 94070, (650) 591-4486.

Roofing Contractors Association of Southern California, 6280 Manchester Blvd, STE 104, Buena Park, CA 90621 (714) 522-4694.

Roofing Industry Council, 400 Reed St, STE D, Santa Clara, CA 95050.

Sacramento Builders Exchange, 1331 T St, Sacramento, CA 95814, (916) 442-8991.

San Francisco Builders Exchange, 850 S. Van Ness AVE, San Francisco, CA 94110, (415) 282-8220.

San Luis Obispo County BLDG Contractors Association, 3563 Sueldo St, STE G, San Luis, CA 93401 (805) 543-7016.

Santa Barbara Contractors Association, PO Box 41622 Santa Barbara, CA 93410, (805) 964-9175.

Santa Maria Valley Contractors Association, 2003 N Preisker LN Santa Maria, CA 93454, (805) 925-1191.

Shasta BLDG Exchange, 2990 Innsbruck, Redding, CA 96003 (530) 221-2140.

Society of American Military Engineers, Orange County Post, c/o Tim Kashuba, Moffatt and Nichol, 250 Wardlow Rd, Long Beach, CA 90807, (213) 426-9551 - FAX: (213) 424-7489.

Southern California Builders Association, 4552 Lincoln, STE 207, Cypress, CA 90630 (714) 995-5841.

Southern California Drywall Contractors Association, 111 Town and Country Rd, STE 45, Orange, CA 92668, (714) 998-8125.

Southern California Environmental Balancing Bureau, PO Box 605, Santa Ynez, CA 93460.

Ventura County Contractors Association, PO Box 7365, Oxnard, CA 93031, (805) 981-8088.

Western Electrical Contractors Association, Sacramento Valley Chapter, 7500 14th AVE #25, Sacramento, CA 95820, (916) 453-0112.

Western States Ceramic Tile Contractors Association, 5004 E. 59th Pl, Maywood, CA 90270, (213) 560-1673.